Science and Technology
on the Internet
PLUS

An Instructional Guide

*Includes diskettes with presentation slides
in PowerPoint for Macintosh or Windows*

Internet Workshop Series Number 4

**Supplement to
*Crossing the Internet Threshold***

Gail P. Clement

Library Solutions Press
Berkeley and San Carlos, California

First printing: June 1995

Graphics Editors: Catherine Dinnean and Stephanie Lipow

Copyright © 1995 Gail P. Clement

LIBRARY SOLUTIONS PRESS

Sales Office: 1100 Industrial Road, Suite 9
                           San Carlos, CA 94070

Fax orders: 415-594-0411

Telephone orders
and inquiries: 510-841-2636

Web URL   http://www.internet-is.com/library/

Email   info@library-solutions.com
                           [leave subject and message blocks blank; you will
                           receive an automated response]

Editorial Office: 2137 Oregon Street, Berkeley, CA 94705

ISBN: 1-882208-13-7

Table of Contents

SECTION C:

Foreword by Anne G. Lipow, Series Coordinator

About this series...

The volumes of the Internet Workshop Series are the actual workshops, in book form, of expert Internet trainers and include their well-tested lectures, demonstrations, exercises, and handouts. The series is, therefore, intended to be useful to two types of reader: the trainer and the learner.

Each title comes in two editions: the "book-alone" edition and the "PLUS" edition. With one exception, the book-alone edition is intended primarily for the learner. That exception is Volume 1, "Introducing the Internet": both versions of the book are intended for the trainer who is preparing an introductory lecture and demonstration—not a hands-on session. The book-alone editions are also useful to trainers as models on which to pattern their own, and some of the titles include the presentation slides in full-size form from which the trainer can make transparencies.

The PLUS edition is intended specifically for the trainer and always includes the presentation slides—in full-size for making transparencies, as well as on disks (Macintosh and Windows) for displaying the slides from a computer. Also, owners who have the PowerPoint software (which was used to create the slides) can customize the slides to suit their own sessions and display or print from a computer.

For the learner, each title is a self-paced workshop-in-a-book. The learner is expected to concentrate on the lecture and overheads and to skip parts addressed to the trainer. However, the lecture and overheads may not be sufficient for understanding the topic because there are two critical pieces of the live workshop that are missing: (a) the online demonstration and (b) the class discussion. To compensate for the absence of online demonstrations, the reader should go through the exercises and handouts systematically. And to experience, though in a delayed fashion, the give-and-take between instructor and student, we urge you to note your questions in the margins and email them to the instructors. (Their email addresses can usually be found at the end of a section in the Preface.) Future printings may be revised to answer such questions for later readers.

For the trainer, each volume provides a model training tool. With the astounding rate of growth of the Internet, it is likely that as soon you learn about an aspect of the Internet, you'll be asked to explain it to others. Of course, knowing a subject is one thing; teaching it so that your audience learns is quite another. That's the primary reason for the PLUS editions: by example to provide the new trainer with the basic skills needed for a successful instructional program. Each volume gives you the words and supporting materials of a proven training session and provides asides to the trainer: for example, advice about how to handle a tricky segment; the principle underlying a particular way of dealing with a topic, the equipment needed for the session. PLUS edition purchasers are welcome to use the lecture and overheads for their own in-person instructional sessions only. Trainers may wish to coordinate bulk orders of the book- alone edition for use as a student workbook, available at a discount from the publisher. For questions about bulk orders or for clarification of copyright issues, please contact us at sales@library-solutions.com

Whether you are a trainer or a learner, your comments about the usefulness of the volume you are using are most welcome. Please address them to the author, whose email address can be found in the Preface to each book.

...and this book

When my friend Patty Iannuzzi, head of the reference department where author Gail Clement is a librarian, said, "Anne, if you want a book in the Internet Workshop Series on Internet resources in the sciences, I know the best person to do it" she was right. There are many things that are special about this book. First, it is not only a helpful tutorial—clear, step-by-step instructions and comprehensive coverage—but it is also a good read for the scientist as well as the lay person. A non-scientist would find this book a wonderful source of information about what the scientist does for a living, as well as about the amazing directions the Internet is taking, especially in the areas of face-to-face meetings, finding information, and sharing resources. (It's a pity that, because of its title, this book is unlikely to be seen by the non-scientist!)

Another thing that makes it special is that you really do not ever need to touch a keyboard to see how you would accomplish a particular task on the Internet. The exercises, with the author's commentary running along-side the screen displays, are quite self-evident.

Third, within six of the seven instructional modules (Section A), under the heading "Quality and Value," are important guidelines for evaluating what you retrieve. I have not seen in any other book about the Internet such attention to evaluation issues. Using criteria for selecting materials is second nature to librarians; it is part of their training. But most of the rest of us do not have explicit yardsticks for judging the validity of the information we've retrieved. While one of the Internet's great strengths is its *de facto* policy of accepting everything that fits, that is also one of its great weaknesses: you can too often and too easily retrieve a mountain of junk. So knowing whether your information is *good* is a most important skill to have when using the Internet. You will be a smarter Internet user for having learned from this book how to cull the gems from the junk.

And not to contradict the author, but although she warns the reader that this book assumes some basic Internet knowledge, I believe that if you only know how your local system works to get you to the various Internet tools (Telnet, Gopher, FTP, WWW, *etc.*), even if you've never used them, you'll easily make your way through this book. Let me know if I'm wrong: write to me at anne@library-solutions.com.

About the Author

Gail Clement has been applying global networking technologies in research settings for over a decade. A former research assistant at NASA's Goddard Space Flight Center, the Smithsonian's Museum of Natural History, and the international Ocean Drilling Program, she has conducted scientific investigations in wide-ranging subjects, from meteorites to marine ridges, and has worked with scientists worldwide in the laboratory, out in the field, and on board ship.

In her current position as Science/Information Services Librarian at Florida International University, Ms. Clement combines her interest in the sciences and her expertise in network applications to develop services and collections that serve the day-to-day needs of faculty, students and the local research community. Since 1992, she has developed and led an ongoing Internet training program, as well as course-related instruction for the sciences and engineering. She serves on the Library's Network Access committee, Vision Group, Systems Group, as well as the University-wide World-Wide Web Committee and the University Research Council.

Outside of FIU, Ms. Clement has designed and led Internet training programs and presentations for librarians and researchers. In the American Library Association, she is a charter member of the Library and Information Technology Association's Internet Room Steering Committee (the group that designs and manages the Internet Room at ALA), and for the last year has served as its chair. She also coordinates the Internet Demo Room for the ONLINE/CD-ROM conference. She writes and speaks frequently on the subject of Internet and its applications for libraries and the research community. Her articles have appeared in *Internet World* and *Database* magazines.

Ms. Clement holds Master of Science in Geology from the University of Oregon and a Master of Library and Information Science from the University of South Florida. She lives in Miami with her husband and two sons.

Preface

About This Guide

The primary objective of this guide is to convey the concepts, skills, and strategies needed to take full advantage of the network's resources for science and technology. It offers a practical approach to meet the everyday needs of people involved in scientific research and practice:

- scientists, engineers, and technical managers, increasingly reliant on the Internet to carry out the business of the day

- librarians and information specialists, challenged to integrate Internet resources into existing collections and services

- Internet trainers, charged with developing effective learning experiences for scientific and technical users *

Each instructional module emphasizes tasks or activities common to sci/tech professionals, demonstrating the Internet tools and resources most useful in accomplishing them. Also covered are criteria for assessing quality and value of information available on the Internet, and strategies for finding appropriate resources in a specific area of interest.

As a supplement to the bestselling Internet guide, *Crossing the Internet Threshold* (Berkeley, CA: Library Solutions Press, 1995), this book is not intended as an Internet manual for new users. Readers will not find in this guide basic information about Internet history and protocols, types of connections, or general 'how-to' information about Internet tools. This type of information is covered quite effectively in *Crossing the Internet Threshold* and numerous other basic Internet guides now available on the shelves of bookstores and libraries.

Rather, this book is designed to serve those with some basic Internet familiarity, interested in using the network in their day-to-day work. To this end, it emphasizes applications most relevant and useful for scientists and engineers today:

> Conferencing and collaboration
>
> Finding colleagues
>
> News and current information
>
> Reference Tools
>
> Searching (and retrieving) the literature
>
> Electronic publishing
>
> Sharing data and primary resources

---

* See the next section of this Preface for a sample workshop outline and suggestions for how a trainer might adapt it for different time frames. The PLUS edition (*Science and Technology on the Internet PLUS*) includes copies of presentation slides on 8.5" x 11" pages for making transparencies to accompany the lecture, and diskettes of those presentation slides for displaying in color or black and white from a computer.

The approach underlying this instructional program recognizes that sci/tech users of the Internet, like any group, have specific needs, interests, and concerns. Their work is, by its very nature, global, interdependent, and collaborative. The information they use or produce is, by necessity, high-quality, authoritative, reliable, and current. And, independent of the Internet, they rely on a number of established tools and resources for scientific research and communication, most of which are still effective and widely-used today.

These characteristics of sci/tech Internet users have shaped the learning objectives promoted in this guide. It is hoped that the reader will:

1. Understand Internet applications for scientific and technical tasks (What can I use the Internet for?)

2. Appreciate the variety and type of sci/tech resources on the Internet (What's Available?)

3. Recognize the differing levels of quality and value of sci/tech resources on the Internet, and develop strategies for assessing them (How useful or reliable is the Internet resource I've found — in its own right, and in comparison with resources available in print or elsewhere?)

4. Use a variety of Internet tools to accomplish sci/tech-related tasks (What skills and access methods are needed to use Internet resources for science and technology?)

5. Develop strategies for keeping up (How do I learn about and locate resources in my areas of interest?)

The Four Sections of This Guide

Section A comprises an entire workshop series, in seven modules, with lecture notes and exercises. Each module stands alone, and may be used in combination with any other module, in any order that makes sense for you.

Each lecture is organized under headings that reflect the issues of greatest concern to sci/tech users of the Internet. In other words, each heading directly addresses the learning objectives described above. (Trainers may use these headings as the basis for presentation slides that accompany the lecture).

The exercises accompanying each module demonstrate how various Internet tools may be applied to solve a problem or fulfill a research need. They also serve to highlight the diversity of resources that might be valuable to researchers in science and technology, and practical strategies for finding and using them. Instructors may use or adapt these exercises for online demos during a lecture, or for practice during a hands-on session.

Section B is a collection of 'Fact Sheets' organized by specific Internet tool (*e.g.*, Telnet, FTP, MOO) For each tool, a summary of scientific/technical applications is provided, with keys to relevant exercises from Section A. Refer to these Fact Sheets for quick summary information about particular tools. Instructors offering tool-based training (*e.g.*, programs emphasizing a particular Internet tool, such as Gopher, FTP, WWW) may consult these sheets for guidance in customizing prepared training materials for sci/tech audiences.

Section C addresses both learners' and instructors' need to find Internet resources to meet a specific purpose or area of interest. This section lists and describes tools, techniques, and other resources most useful in identifying and locating Internet resources across the diverse areas of science and technology. Instructors may use these sites to select those of interest to their audience, or may offer some free time during the worshop for participants to try different finding techniques.

Section D is a selective list of articles, reports, and other documents which will help in understanding how the Internet is being used in the many disciplines of science and technology, and its impact on the processes of scientific investigation and communication. These materials may provide background material or pointers to interesting sites and applications for both learner and instructor.

How to Follow the Exercises

The exercises included in each instructional module of Section A demonstrate the types of tools and resources most useful for sci/tech applications on the Internet. They demonstrate a small but representative sampling of the many Internet resources and sites within the pure and applied sciences.

All of the exercises included in this book have 'happy endings.' In practice, however, snags do occur. Many of the pitfalls and problems you might experience are noted in the annotations to the exercises, in the lectures, and in the Fact Sheets for specific Internet tools provided in Section B.

For the sake of consistency, the exercises represent actual sessions on the UNIX operating system, using common UNIX software such as the curses Gopher client, the PINE mail program, and the text-based Lynx Web browser, *etc.* These choices in no way suggest that the reader must use these specific software programs to perform the same tasks. All of the exercises provided in Section A may be executed with similar effect from any type of operating system, and from most every Internet software program.

To help the reader distinguish between text typed in by the user, and text displayed by the computer system, the exercises have been formatted with various fonts. Text that appears in `courier` font denotes text displayed on the computer. **`Boldface courier`** font denotes text entered by the user. Text appearing in ***`boldface courier italics`*** indicates a generic term that the user will replace with a specific term, such as a user ID, filename, etc.

In most cases, the exercises include screen displays from actual sessions on the Internet. To save space, long displays were truncated or text omitted. In any exercise involving personal usernames or email addresses, including all those prepared for the "Finding Colleagues" module, actual content was edited in order to protect the privacy of individual Internet users.

All exercises were verified at the time this book went to press, but the access method, address, content, or arrangement of any site is subject to change at any time. Readers are invited to forward any corrections, updates or additions by electronic mail to clementg@servms.fiu.edu.

To the Instructor: Putting This Guide to Work

A Trainer's Sample Workshop Outline

Workshop Title: Science and Technology on the Internet: An Overview

Format: One-time workshop

Running time: 1³/4 hours (lecture only) to 8 hours (lecture, demos and exercises)

Audience: Users with basic understanding of Internet history, concepts, vocabulary; experience with at least common tools (email, ftp, telnet, gopher, www)

Outline for 4-hour version of workshop

Introduction and Icebreaker

Welcome Remarks

Introduce Instructor

Introduce Participants

Go over Workshop Outline and Objectives

Go over Workshop Materials

- Lecture
- Structured Group Exercise
- Questions and Answers

Science, Technology and the Internet: A Little Background

Relevant Sections: Preface and Section D

History of Internet in Scientific/Research Community

Impact of Internet on Science and Technology

Current and Emerging Trends

Future Prospects

- Lecture with Slides
- Questions and Answers

What are the primary issues of concern for sci/tech users of the Internet? (And Why Are They Important?)

Relevant Section: Section A

What can I use the Internet for?

How useful or reliable are Internet resources — in their own right, and in comparison with resources available in print or elsewhere?

What skills and access methods are needed to use sci/tech Internet resources?

How do I learn about and locate resources in my areas of interest?

- Lecture with slides
- Demos (live or canned)

Sci/Tech Applications of the Internet   *Relevant Section: A*

Finding Colleagues

Conferencing and Collaborating

News and Current Information

Reference Tools

Searching (And Retrieving the Literature)

Network Publishing

Data Exchange and Resource Sharing

Add/substitute other applications important for that group
(Instruction, marketing, etc.)

- Lecture with bullet points
- Demo of each application
  (live or canned—may use exercises from Section A)

Wrap-up   *Relevant Sections: B, C, D*

Where and how to learn more?

Workshop Evaluation

- Lecture
- Group Discussion
- Questions and Answers

Adapting workshop for differing formats

Shorter timeslot:

Lecture only; reduce demos

Select fewer applications from Section A to cover

Longer timeslot:

Add demos

Add selected hands-on exercises from Section A

Add 'free-time' to find sites of interest from Section C

Continuing Series:

Offer overview session (described above) as first workshop;
cover different applications from Section A each week thereafter

Assign exercises for interval between workshops

Science, Technology and the Internet: A Little Background

From its inception, the Internet has been a tool for scientists and engineers. The early ARPANET of the 1960's and 1970's was the vision of computer researchers interested in connecting their machines in order to share ideas, data, and other resources. The success of their early experiments catalyzed a revolution in science computing, as more scientists and engineers joined the network to share information, collaborate on papers, and participate in online discussion.

The subsequent expansion of the network's backbone by the National Science Foundation in the 1980's and early 1990's furthered the Internet's role as an essential tool and resource for scientific and scholarly communication and research. Network access spread to scientists, engineers, librarians, educators, and others in academic and research centers worldwide. Now network access is nearly ubiquitous among the scientific and research community. And, not surprisingly, the last few years have seen the explosive growth of Internet applications for most every scientific and technical field.

In assessing the impact of the Internet on science and technology, it is important to bear in mind that scientists engaged in global networking long before the Internet. Research success has always depended on close communication and resource sharing with colleagues around the world. Scientists have always been skilled in retrieving data from distributed sources worldwide and synthesizing them into a logical whole. And, for hundreds of years, scientists have been communicating their ideas and knowledge—in the form of a scientific paper, proposal, or presentation—to a forum of peers for comment, critical review, questioning and judgement.

Before the emergence of worldwide computer internetworks, scientists accomplished these tasks by applying the technologies of the day: postal delivery, phone, fax, printed media such as the science journal, and transportation to join colleagues in the lab, field or conference. Now they use electronic mail, discussion groups, File Transfer Protocol, Telnet, Gopher, and the World-Wide Web to accomplish the same. It is evident, then, that the Internet has automated many of the processes of research and communication. In other words, the Internet has enabled scientists to perform ordinary tasks more efficiently, quickly and effectively .

In various sectors of the scientific community, there are also signs that the Internet is enabling scientists to perform tasks in entirely new ways, and to perform entirely new tasks. During events surrounding the collision of comet Shoemaker-Levy with planet Jupiter, the Internet facilitated unprecedented coordination of observation and data-sharing efforts, demonstrating its value as a tool of collaboration and communication. The biologists' BioMoo, an online research center, illustrates the power of interacting with colleagues worldwide through virtual laboratories. Employing Internet's multicast backbone, or MBone, Woods Hole Oceanographic Institution transmits data and real-time video from their remote vehicle system Jason/Medea to scientists' desktops on shore and to dozen of classrooms across the country. The *Journal of Artificial Intelligence Research* and the *Electronic Journal of Combinatorics*, two of the few dozen peer-reviewed science journals published exclusively on the Internet, demonstrate the effectiveness of hyperlinked pages to draw together multimedia resources distributed worldwide. These are but a few of the 'revolutionary' scientific applications emerging on the Internet today.

The tremendous potential of the Internet not only to automate, but to transform, has led visionaries to conceive new notions of how scientific communities might function in the future. Most compelling is the concept of a shared workspace or "collaboratory"

> —a center without walls, in which researchers can perform their research without regard to geographical location—interacting with colleagues, accessing instrumentation, sharing data, and computational resources, and accessing information in digital libraries.[1]

In some corners of the scientific community, we are already seeing the first steps toward realization of this vision. The prototype Worm Community System,[2] developed for researchers in molecular biology, and new collaboration testbeds in upper atmospheric science,[3] engineering,[4] and in various other fields, demonstrate the capabilities of global networking to advance team science, facilitate resource sharing, and enhance access to both informal and formal communications.

Expanding the success of these early experiments in the greater scientific community will take the fuller understanding and wider acceptance of existing network tools and resources to perform day-to-day tasks. It will also require the development of new tools to accomplish new tasks—visualization, tele-experimentation, videoconferencing, and others—not widely available with current network technologies.

Ultimately, visionaries predict that we will see the emergence of new electronic communities of scientists and engineers, in which the dispersed members and myriad resources of a given profession will be immediately available from the desktop.[5] Moreover, the new tools and resources that develop to meet the needs of scientists and engineers will no doubt find application in other fields and for other purposes, benefiting not only researchers, but the educational community, the marketplace, and society as a whole.

The scientific community was first to recognize the power and promise of the Internet. It is certain to lead its future development.

Endnotes

1 The collaboratory concept was originally developed by Dr. William A. Wulf and presented in a white paper entitled "The National Collaboratory -A White Paper" which was included in an unpublished report from an invitational workshop, "Towards a National Collaboratory" held in March 1989 at Rockefeller University.

   The quote included here was excerpted from a subsequent publication of the National Research Council:

   Commission on Physical Sciences, Mathematics, & Applications Staff, National Research Council. *National Collaboratories: Applying Information Technology For Scientific Research.* Washington D.C.: National Academy Press, 1993, p. vii.

2 Pool, Robert. "Networking the Worm" *Science* vol. 261, no. 5123 (August 13, 1993): 842.

3 Information about the "Upper Atmospheric Research Collaboratory (UARC)", under development by the University of Michigan and several other organizations, is available from the university's School of Information and Library Studies at:

   `http://http2.sils.umichigan.edu/UARC/HomePage.html`

4 Information about the project "Building the Interspace: Digital Library Infrastructure for a University Engineering Community" under development by the University of Illinois and other organizations, is available from the university's Grainger Engineering Library at:

   `http://www.grainger.uiuc.edu/dli`

5 Pool, Robert "Beyond Databases and E-mail" *Science* vol. 261, no. 5123 (August 13, 1993): 841, 843

Acknowledgements

This book, and the instructional programs it represents, could not be accomplished without the support and encouragement of many colleagues, friends, and family members. I wish to thank Anne Lipow for her unending patience, support, and constructive comments, and her staff at Library Solutions for their creative approach and eagle eyes.

Thanks are also due to Thomas Dowling, Sharyn Ladner, and Linda Pikula for reviewing the manuscript. Dr. Abdesselam Bouzerdoum, Department of Electrical & Electronic Engineering at the University of Adelaide, Australia, graciously permitted us to reprint a portion of his article from *Complexity International* in the "Electronic Publishing" module (Publishing Exercise 7).

I wish to acknowledge Patricia Iannuzzi, Head of Reference at the Florida International University Library, for facilitating the meeting of publisher and author, and the rest of the staff at FIU for their encouragement and interest. Many other faculty, students, and other members of the FIU community have also provided motivation, feedback, and helpful suggestions over the past few years.

To Norma P. O'Haire go my thanks for her unwavering interest and support. And finally, I want to express very special appreciation and much, much more to Bradford Clement, who always believed, and to Tyler and Peter Clement, for giving up time with the computer (and many other things) so that I could finish the book.

clementg@servms.fiu.edu
June 1995

Section A: Instructional Modules

Throughout this section are numbered projector icons that correspond with the numbered presentation slides in Section E. A slide should be displayed when you reach its corresponding number in the text.

The diskettes at the back of the book contain the same slides in color. The Readme file on each diskette will explain how to display the slides.

 Instructor's preliminary remarks:

Welcome, Introductions, Expectations

 Module 1: Conferencing and Collaboration

Module 2: Finding Colleagues

Module 3: News and Current Information

Module 4: Reference Tools

Module 5: Searching (and Retrieving) the Literature

Module 6: Electronic Publishing

Module 7: Sharing Data and Resources

Module 1: Conferencing and Collaboration

IN THIS MODULE:

Introduction
What's available
Quality and value
How to find what's available
Exercises:

Introduction

The practice of conferencing and collaborating with colleagues world-wide is nothing new for scientific researchers. Traditional forums for group discussion and interaction include common membership in scientific societies, participation in conferences and workshops, joint ventures in laboratory or field, shared service on editorial boards or review panels, and so on. Participation in these forums may require travel, long-distance phone calls, or the sluggish delivery of messages and materials through postal service and other delivery systems.

The Internet offers global scientific communities a group of new applications for conferencing and collaboration: electronic discussion groups, virtual environments, videoconferencing, and emergent tools for shared whiteboards, visualization, and more. These online communication forums enable researchers to interact with colleagues immediately, without leaving the office or laboratory. Their advantages include not only savings of time and expense, but also the opportunity to reach colleagues in small or underdeveloped nations, otherwise isolated by pecuniary or political factors.

There is also increasing recognition that communications systems based on the Internet have the potential not only to automate and expand scientific interaction, but to fundamentally change it. Visionaries observe that Internet offers an unprecedented level of interactive feedback on a global scale, dramatically enhancing the processes of scientific inquiry and communication. Steven Harnard, editor of the network ejournal *Psycoloquy*, has predicted that electronic conferencing or, in his words, "scholarly skywriting," "will substantially restructure the pursuit of knowledge."[1] William A. Wulf, writing about a future "collaboratory," or research "center without walls" emphasizes that "... fuller exploitation of the power of information technology can not only increase the productivity of our researchers, it can qualitatively change the kinds of questions we ask and, hence, what we know about nature...."[2] And Bruce Schatz, developer of the Worm Community System—a successful prototype collaboration environment for molecular biologists—predicts that the new electronic communities forged by network technologies will revolutionize scientific research by providing a new way for scientists to "record and share information and insights."[3]

How the Internet will prove to revolutionize the process of scientific communication remains to be seen. What is evident is that electronic conferences on the Internet have become testbeds for new ideas, meeting grounds for new partnerships, and the shared workplaces of collaborating colleagues. In short, they have become indispensable communication tools for their given areas of specialty, enabling researchers and other sci/tech professionals to carry out the business of the day more easily and effectively.

 3

WHAT'S AVAILABLE

Whether you are interested in joining an existing conference online, or starting up a new one, the Internet offers a variety of applications for interaction among colleagues in dispersed locations. Their capabilities range from deferred messaging in text-only format, to the immediate exchange of graphics, data, and other multimedia, or face-to-face discourse transmitted in real-time.

Electronic Discussion Groups

Why use them?

The most popular venue for conferencing and collaboration on the Internet are the electronic discussion groups, ubiquitous because they are relatively easy to set-up, maintain, and use. These groups—also known as conferences, discussions lists, or newsgroups—present discussion as text-based contributions that are submitted by one individual and almost immediately transmitted to all other members of the group. Representing online communities with a common interest, these groups attract a wide range of subscribers, from senior researchers and technical managers to graduate students and others new to the field.

Electronic discussion groups tend to focus on narrow research areas that parallel the subspecialities found in most every scientific and technical field. The content of what is submitted to these groups can be thought of as conversations with colleagues and can range from intellectually substantive to practical and immediately useful.

Electronic discussion groups may also serve as natural clearinghouses for news, announcements. Some groups, in fact, are established only for this purpose. Subscribers routinely receive announcements concerning upcoming conferences, professional vacancies, opportunities for funding or fellowships; tables of contents for journals; reviews of books, software, *etc.*

Electronic discussion groups may be moderated, meaning that articles submitted to the group are first approved by someone on the basis of relevance (and in some groups, on the quality of the content). Many sci-tech discussion groups are not moderated, to allow discourse to flow more freely and quickly. In either case, the topics covered by the group are usually specified in the welcome letter sent out to new subscribers, or in a Frequently Asked Questions (FAQ) document compiled for the group.

 4

Access Methods

Electronic discussion groups take two principal forms: mailserver lists, distributing articles directly to subscriber's electronic mailboxes, or Network Newsgroups, distributing articles to central news servers for users to log into and read. Which format a particular group uses is somewhat arbitrary, relating to the age of the group and who started it. Many groups now run in parallel through both a mailserver and on Network News (*e.g.*, `bit.listserv.frac-l` and `frac-l@gitvm1.gatech.edu`), offering members the option to follow and participate in group discussion with either access method. Those preferring to keep their electronic mail boxes free from unnecessary or unwanted messages may prefer to follow the group via Network News. Others may prefer to subscribe via mailserver, to be sure of automatically receiving a copy of every message posted to the group.

The principal features, capabilities, and commands of both mailserver programs and the Network News system are detailed in the corresponding Fact Sheets in Section B of this guide.

To access the backfiles, or archives, from previous discussions, or the FAQs, "help" documents, and other files related to a particular group, you may need to submit specific commands to the relevant mailserver, or use one or more popular search and retrieval tools such as Gopher, WAIS, and FTP. The specific methods for accessing files related to a particular conference or group are usually explained in its welcome letter, introduction file, or FAQ document.

Virtual Environments

Why Use Them?

Another application for conferencing and collaboration gaining wider use on the Internet is the "virtual" environment, simulating the real-life laboratories and buildings populated by researchers and other sci/tech professionals. These forums for group discussion may take the form of chat services, with participants gathering on one more channels dedicated to a specified theme; or the richer environment of the MOO (Multiuser Dimension, Object-oriented), offering opportunities to interact not only with other people, but with objects as well.

Though originally (and still widely) used for recreational purposes on the Internet, these venues have been adapted for members of the scientific community with the creation of familiar workplaces—laboratories, offices, seminar rooms, lounges, and more. Visitors and dwellers in these environments move about, search for colleagues, read news, scan bulletin boards, and make conversation by typing in text commands from the keyboard.

Examples of virtual environments developed for scientific use include EnviroLink's EnviroChat service, available most any time for discussions about environmental issues[4]; the University of Virginia's EcoChat electronic conference room,[5] and the Weizmann Institute's BioMOO,[6] providing a complex research center that is 'virtual' home to numerous researchers in bioscience.

Though originally developed (and still mostly limited) as text-based communications systems, some virtual environments are being adapted to provide expanded capabilities for graphics and other multimedia expression. But even the most 'primitive' environments foster effective interactions with people or objects—using commands for conversing, emoting, reading, scanning, and so on, researchers carry on everyday work in laboratories, lounges, journal clubs, and other workspaces constructed only with simple text.

Access Methods

Access to a virtual environment is usually available through Telnet connection, either directly or through a Gopher or World-Wide Web gateway.

The software programs underlying each of these environments may use comparable, but slightly different, commands so it is best to get a list of available commands by invoking online help upon entering the system. For more information about using Internet Relay Chat or MOO, the software tools underlying most virtual environments, see the corresponding Fact Sheets in Section B of this guide.

Files related to the various virtual environments are sometimes available for searching and retrieving outside the environment itself. For example, Wiezmann Institute's BioMOO maintains a 'FAQ,' some help files, archives of journal club gatherings and other meetings, on its Gopher site

```
gopher://bioinformatics.weizmann.ac.il:70/11s/biomoo
```

Videoconferencing and Multicasting

Why Use Them?

Among the most promising Internet applications for conferencing and collaboration are the developments in videoconferencing or multicasting that relay audio and/or video in real-time to desktops around the world. The ability to carry on immediate face-to-face discussion with these tools comes closest to simulating the type of interaction one experiences in 'real life' conference environments.

The remote workshops offered by the National Center for Atmospheric Research for various scientific organizations,[7] and the now routine live audiocasts of meetings of the Internet Engineering Task Force (IETF),[8] demonstrate the tremendous potential of Internet videoconferencing for distance learning, teleconferencing, and other collaborative endeavors.

Access Methods

Videoconferencing or multicasting services, such as MBone (pronounced em-bone), C U C Me (pronounced see-you-see-me), and many other tools are now in various stages of experimentation or development. The sophisticated operating requirements of these real-time, multimedia communications services (e.g., specialized software, direct Internet connections with high bandwidth, specialized hardware and other equipment), prohibit access by most users today. However, their powerful capabilities are attracting much interest within the scientific community, and work is well underway to making these tools more widely available.

For more information on MBone, the multicasting application already in use by many scientists and engineers, see the corresponding Fact Sheet in Section B of this guide. To learn more about C U C Me, you may refer to the C U C Me FAQ available at:

```
ftp://ftp.galed.cornell.edu/pub/video
```

Collaboration Environments

Why use them?

Collaboration environments combine the capabilities of discussion, instrumentation, data analysis and manipulation, and information search and retrieval, into a virtual research center, bringing to each researcher's desktop all the resources needed to perform day-to-day work.

Although visions of a collaboration environment have been described for many years, efforts to actually develop one did not advance until global computer networks gained widespread application within the scientific community. The Internet now offers sufficient hardware and software to support endeavors to fulfill this vision.

One of the first collaboration environments successfully developed for scientists was the Worm Community System (WCS), introduced by University of Illinois scientist Bruce Schatz as a prototype community knowledge base for biologists studying the model organism *C. Elegans*, a nematode worm. As a combined electronic communication tool and database, the WCS integrates both the informal and formal knowledge of a community.[9] Spurred by the early success of the WCS and by initiatives of the United States' National Research Council and National Science Foundation, new collaboration testbeds have also begun in upper atmospheric science, in space science, and in other fields.[10]

Access Methods

Emerging technologies to support collaboration over the Internet are in various stages of development or experimentation. Information about a number of such tools is available from the Software Development Group at the National Center for Supercomputing Applications (NCSA). A visit to their World-Wide Web site provides descriptions, demonstration, and in some cases, downloadable versions of new software developed for scientific collaboration:

<code>http://www.ncsa.uiuc.edu/SDG/SDGIntro.html</code>

One of the next generation tools now available for downloading and testing from NCSA is Collage, a type of scientific "groupware" that enables collaborating colleagues to analyze and visualize data, edit shared texts, and exchange ideas and information. Collage is a graphical application with versions available (at this time of writing) for Unix and Macintosh.

 5

QUALITY AND VALUE

As forums for informal discussion and interaction, Internet conferences generally do not meet the quality control standards characterizing more formal scientific communications (*e.g.*, peer reviewed journals) Nevertheless, certain filtering mechanisms may be in place to ensure that discourse is, at a minimum, relevant to the group (and, in some cases, sufficiently substantive as well).

Screening of participants

Some groups are open only to select participants. Others may screen subscription requests. Most virtual environments offer 'conference rooms' for advance reservation, or private channels for exclusive groups.

Moderated discussion

Electronic conferences that operate with moderators tend to eliminate the most grievous subscriber violations because one member of the group routinely filters articles based on content. In some cases, the moderator may also serve as discussion leader, initiating new 'threads' for discourse, or as referee, closing or redirecting discussion that goes on too long, gets off-track, or needs to be continued privately, outside the group.

In virtual environments, a long-term resident may serve as 'unofficial' moderator and host, approaching silent lurkers with offers of help or guidance, or shepherding the wayward novice to a 'safe' starting point.

Self-regulation

By and large, it is through self-regulation or "good netiquette" that most electronic groups maintain the integrity of their discussions. New members learn the appropriate scope and behavior for the group by reading its welcome letter or FAQ. Contributors deemed 'out-of-line' are usually so informed immediately by the heated responses (also known as 'flames') of fellow subscribers. In extreme cases, a wayward group member may be unsubscribed from or asked to leave the group.

HOW TO FIND WHAT'S AVAILABLE IN YOUR FIELD

Finding an electronic discussion group on your subject is fairly easy, thanks to a number of directories developed just for this purpose. Those organized by subject include:

"Dartmouth Special Internet Group List (Siglist)"
`http://alpha.acast.nova.edu/cgi-bin/lists`

Hanford and Buckman's "LISTSERV Information Home Page"
`http://www.clark.net/pub/listserv/lsv11.html`

Kovacs et al.'s "Directory of Scholarly Electronic Conferences"
`gopher://gopher.usask.ca/11/Computing/Internet20%Information/`
`Directory%20of%20Scholarly%20Electronic20%Conferences`

Vivian Neou's "List of Lists"
`gopher://crvax.sri.com:70/00GOPHER-ROOT1:[NETINFO]interest-`
`groups`

For additional information on finding discussion lists of interest, see:

Arno Wouters, "How to Find an Interesting Mailing List" (February 1994)

```
mailto://listserv@vm1/nodak.edu/get new-list wouters
```

```
ftp://vm1.nodak.edu/new-list/new-list.wouters
```

Finding topical virtual environments, videoconferences, and multicast events for real-time interaction with colleagues is more difficult. They may turn up on the menus of subject hubs or other resources covering your area of interest. See Section C of this guide for more specific information on finding subject-oriented resources of all types.

Endnotes

1. Harnard, Stevan, "Scholarly Skywriting and the Prepublication Continuum of Scientific Inquiry." *Psychological Science*, vol. 1 (1990): 342-343.

2. Wulf, William A. "The Collaboratory Opportunity." *Science*, vol. 261, no. 2153 (August 13, 1993): 854.

3. Pool, Robert, "Networking the Worm." *Science*, vol. 261, no. 2153 (August 13, 1993): 842.

4. To reach EnviroChat, telnet to

```
envirolink.org, port 2000.
```

5. EcoChat is a project of the University of Virginia Office of Recycling and Environmental Information. To get information about reserving EcoChat for a conference, see the EcoChat on the University's EcoGopher

```
gopher://ecosys.drdr.Virginia.EDU:70/00/chat/EcoChat
```

6. For more information on BioMOO, see the informational files on the Weizmann Institute of Science's Bioinformatics gopher

```
gopher://bioinformatics.weizmann.ac.il:70/11s/biomoo
```

7. McCann, Mike, "MBone and Distance Learning at the Naval Postgraduate School". *Computer*, April 1994: 31.

8. Eriksson, Hans, "MBONE: The Multicast Backbone." *Communications of the ACM,* vol. 37, no. 8: 54-59.

9. The Worm Community System is described in the article referenced in Endnote 3. Details of the project are also specified in the following NSF Proposal Abstracts: Bruce Schatz, "Designing a Computer System to Support an Electronic Scientific Community" [May 1990 NSF Award 9003540] and Bruce Schatz, "Building an Electronic Scientific Community" [March 1994 NSF Award 9319844] Both abstracts are available from the NSF WWW site at:

```
http://www.nsf.gov
```

10. A number of testbed projects are discussed in the following report: Commission on Physical Sciences, Mathematics, & Applications Staff, National Research Council, National Collaboratories: Applying Information Technology for Scientific Research. Washington, D.C.: National Academy Press, 1993

Conferencing Exercise 1. FIND A DISCUSSION GROUP

Task Find a discussion group on fish ecology. You don't know what type of group it is or where it is hosted.

Approach   Start with one of the good directories of discussion groups, such as the Directory of Scholarly Electronic Conferences and the SRI List of Lists. If one has a relevant listing, a search in the archives of NEW-LIST might turn up something.

Start at the University of Saskatechwan Gopher, which has made the "Directory of Scholarly Conferences" avilable in searchable format.

```
% gopher  gopher.usask.ca
--> Computing/
      --> Internet Information/
          --> Directory of Scholarly Electronic Conferences/
              --> Search Directory of Scholarly ... <?>


Words to search for:  fish ecology

Search Directory of Scholarly ....: fish ecology
```

No suitable listing is found. Perhaps the list developed after this annual directory was last updated.

```
LN: sci.bio.ecology TI: Discussion of various aspects of    ...
LN: ETHOLOGY TI: Discusses animal behavior and behavioral    ...
LN: sci.bio.ethology TI: Animal behaviour and behavioral    ...
LN: ECONET TI: Ecology E-conferencing System (fee-based    ...
LN: ENVBEH-L TI: Environment, Design, and Human Behavior    ...
LN: ECOLOG-L TI: The discussion list for the Ecological Soci...
    .
    .     <display deleted>
    .
```

Now on to the "SRI List of Lists".

It is available as a single, VERY LARGE (932 KB) file using Gopher.

```
% gopher  crvax.sri.com
--> NetInfo/
      --> interest-groups. [30-Nov-1994, 932KB].

interest-groups. [30-Nov-1994, 932KB] (903k)
+-------------------------------------------------------------+
SRI.COM:[ANONYMOUS.NETINFO]INTEREST-GROUPS.TXT[30 Nov 1994]

Welcome to the "List of lists." This is a listing of special
interest group mailing lists available on the Internet.  It was
started by Rich Zellich back when there was only an Arpanet,
and is probably the oldest of the various "list of lists"
available around the Internet.  If you are putting up a copy of
this list in a public place, please let me know so that I can
let you know when new versions come out. If you see ...
+-------------------------------------------------------------+
Search text for:
```

Once Gopher has retrieved and displayed the file, you can start a keyword search of the file with the '/' key.

```
        fish ecology
```

Unfortunately, the fish ecology group is not included in the SRI List either.

With the grit of any seasoned Internet user, you push on to the archives of NEW-LIST@NDSUVM1, to see if you can retrieve the original announcement circulated about the group upon its inception.

Since NEW-LIST@NDSUVM1 is a dicussion group operated by listserv software on a BITNET node, you must use electronic mail to send a search request to the listserver.

If your Internet host computer is not also a BITNET node, you may need to send the mail message through an Internet/BITNET gateway, as shown here.

Instructions for setting up a listserv search are available by sending the message "INFO DATABASE" to the listserver.

After sending a search request to the listserver, you will receive one or more mail messages in return.

The first one reports on the 'output of your job' -- i.e., the status of the search request. If it was successful, a second message will then arrive with the actual search results.

```
are putting up a copy of+--------Gopher Error---------+lease
know so that I can let y|                             |out. If
a+----------------------|   Couldn't find text        |-------+
| Search text for:      |                             |
|                       |   [Cancel: ^G] [OK: Enter]  |
|                       |                             |
|  fish ecology         +-----------------------------+
.
.
.
```

```
PINE 3.91    COMPOSE MESSAGE    Folder: INBOX  44 Messages

To        : listserv%ndsuvm1.bitnet@cunyvm.cuny.edu
Cc        :
Attchmnt:
Subject :
----- Message Text -----
//
database  search  dd=rules
//rules dd *
search  fish  ecology  in  new-list
print
/ *
```

```
PINE 3.91    FOLDER INDEX       Folder: INBOX


N 50   Jan  7 L-Soft listserver (1,038) Output of your job
N 51   Jan  7 L-Soft listserver (4,640) File: "DATABASE OUTPUT"

? Help       M Main Menu  P PrevMsg    - PrevPage     D Delete
```

```
PINE 3.91    MESSAGE TEXT       Folder: INBOX  Message 50 of 51

Date: Sat, 7 Jan 1995 09:52:28 -0600
From: "L-Soft list server at NDSUVM1 (1.8a)"
<LISTSERV@VM1.NODAK.EDU>
To: clement@SOLIX.FIU.EDU
Subject: Output of your job "clement"
```

Here is a copy of the status report from the listserver.

The search was successful.

The actual search results are included in the second of the two messages received from the listserver.

Included in the search results is the original announcement for the group "FISH-ECOLOGY@ SEARN.SUNET.SE".

```
> database search dd=rules
File "DATABASE OUTPUT" has been mailed to you under separate
cover.
```

PINE 3.91   MESSAGE TEXT   Folder: INBOX   Message 51 of 51

```
Date: Sat, 7 Jan 1995 09:52:28 -0600
From: "L-Soft list server at NDSUVM1 (1.8a)"
<LISTSERV@VM1.NODAK.EDU>
To: clement@SOLIX.FIU.EDU
Subject: File: "DATABASE OUTPUT"

> search fish ecology in new-list
--> Database NEW-LIST, 2 hits.

> print
>>> Item number 1673, dated 94/02/15 16:44:55 -- ALL
Date:          Tue, 15 Feb 1994 16:44:55 CST
Reply-To:      SOLARIS@cicei.ulpgc.es
Sender:        NEW-LIST - New List Announcements <NEW-
LIST@NDSUVM1.BITNET>
From:          Aldo-Pier Solari <SOLARIS@cicei.ulpgc.es>
Organization: Universidad de Las Palmas de G.C.
Subject:       NEW: FISH-ECOLOGY - Fish and Fisheries Ecology

FISH-ECOLOGY on LISTSERV@SEARN.SUNET.SE
            or LISTSERV@SEARN.BITNET

    FISH-ECOLOGY is an international computer conference for
academic personnel & students involved in empirical and
theoretical issues related to fish and fisheries ecology:
Evolutionary aspects, population dynamics, modelling,
management, conservation, bioeconomics, related software &
hardware, reviews, symposium announcements, etc.  Membership is
open to all interested parties. Commercial announcements are,
however, not desired.

    The list aims to stimulate connections between senior and
junior researchers and students on an international and
multidisciplinary basis, to exchange views, data and to put
forward ideas to approach fisheries ecological issues.


.
.       <text deleted>
.
```

Conferencing Exercise 2. SUBSCRIBE TO A LISTSERV DISCUSSION GROUP

Task   Subscribe to the electronic discussion group of your choice (shown here are the steps for the listserv discussion group "`fishecology@searn.sunet.se`" found in the previous exercise).

Approach   The subscription procedure you will use depends on the type of group you have selected. If it is a Usenet Newsgroup, you need to read its articles with a newsreader program on your Internet host. If it is a mailserver-based group, you will need to email a subscription request to the computer managing the group (not to the group itself).

Caution! Watch where you send your messages! Messages regarding *subscription* (subscribe, unsubscribe, etc.) should be sent to the *listserver* that manages the list, e.g. 'listserv@searn.sunet.se'. Your *topical* messages should be sent to the group of *subscribers*, e.g., 'fish-ecology@searn.sunet.se'.

To subscribe to the group 'fish-ecology', send an email message to the listserv managing the group.

You will soon receive a response back from the listserver, acknowleding your request.

You will also receive a 'Welcome' letter, summarizing subscriber options and other information about the disucssion group you have joined. Sometimes it even includes a Frequently-Asked-Questions (FAQ) document for the group.

SAVE THIS LETTER! You will no doubt need to refer to it later to find out, for example, how to hold all messages while you are away, or how to unsubscribe from the list.

```
PINE 3.91   COMPOSE MESSAGE   Folder: INBOX  44 Messages

To       : listserv@searn.sunet.se
Cc       :
Attchmnt :
Subject  :
----- Message Text -----
subscribe fish-ecology Gail Clement
```

```
PINE 3.91   MESSAGE TEXT       Folder: INBOX  Message 47
of 47  3%

Date: Sat, 7 Jan 1995 19:19:34 +0100
From: "L-Soft list server at SEARN (1.8a)"
<LISTSERV@SEARN.SUNET.SE>
Reply to: FISH-ECOLOGY-Request@SEARN.SUNET.SE
To: Gail Clement <clementg@SOLIX.FIU.EDU>
Subject: Your request to subscribe to the FISH-ECOLOGY list

Sat, 7 Jan 1995 19:19:34

This document (referred to as FISH-ECOLOGYs frequently
asked questions or FISH-ECOLOGY-FAQ) discusses some questions
and  answers which often arise among subscribers.

   *    List of contents
   *
   * 1. What is FISH-ECOLOGY ?
   * 2. What topics are covered by FISH-ECOLOGY ?
   * 3. What aims does FISH-ECOLOGY have ?
   * 4. What rules do apply in FISH-ECOLOGY.
   * 5. How are messages sent to the entire network ?
```

If you don't save it and find
you need it later, try:
  (1) searching for it in the
group's archives
  (2) sending a message to
the list moderator
  (3) unsubscribe and
resubscribe to the group

Sometimes, you may
receive an initial message
from the listserv asking you
to confirm your subscription
with a statement that you
agree to the group's "Aims
& Rules" . This is one
mechanism for quality
control for the group.

To confirm your request to
subscribe to the group,
choose the '**reply**' option
in your mail program while
the original message is
open and on the screen.

In this case, it is necessary
to include the part of the
original message containing
the Aims & Rules specified
in the FAQ file supplied by
the listserv. All the rest of
the text in that file may be
edited out.

Also, be sure to send your
reply to the listserv (list
manager), not to the list
(subscribers).

Once you reply, your name
is added to the list and you

.
.   <text deleted>
.

* *
                        FISH-ECOLOGY
        Network on Fisheries ecology for professionals
          Aims & Rules: Frequently Asked Questions
                        September 1994
* *
Dear Madame/Sir

In order to confirm your wish to join FISH-ECOLOGY, you
may reply to this message stating your agreement with the Aims
& Rules specified in this FAQ file (Sections 3 & 4). To do so,
you may either use the automatic reply of your mail message
stating your agreement with the Aims & Rules specified in this
FAQ file (Sections 3 & 4). To do so, you may either use the
automatic reply of your mail software or address the message
to:
                FISH-ECOLOGY-REQUEST@SEARN.SUNET.SE

The FISH-ECOLOGY management will be glad to add your entry
as soon as your agreement-and-confirmation message is
received. Thank you.
* *

PINE 3.91   COMPOSE MESSAGE REPLY   Folder: INBOX   47 Messages

To   : **FISH-ECOLOGY-Request@SEARN.SUNET.SE**
Cc   :
Attchmnt:
Subject : Re: Your request to subscribe to the FISH-ECOLOGY
list
----- Message Text -----

Include original message in Reply? **Y**

Use "Reply to:" address instead of "From:" address? **Y**

are free to follow or participate in the group's discussions as you wish.

To send a posting to a listserv group such as 'fish-ecology', email your message to the group itself.

It will then be copied to all subscribers in the group.

You will probably not receive a copy of your own posting to the group. That is the default setting for most listserv groups.

If you do wish to receive copies of your own postings, you need to change your subscriber options by sending a message to the listserv.

For a complete listing of subscriber options available with listserv discussion groups, send a message to the listserv with the subject line blank and 'INFO DATABASE' in the message field.

```
PINE 3.91    COMPOSE MESSAGE REPLY    Folder: INBOX  47 Messages

To         : fish-ecology@searn.sunet.se
Cc         :
Attchmnt:
Subject : Measurement of diameter, salmon eggs
----- Message Text -----
Does  anyone  on  the  list  have  experience  measuring...?
```

```
PINE 3.91    COMPOSE MESSAGE REPLY    Folder: INBOX  47 Messages

To         : listserv@searn.sunet.se
Cc         :
Attchmnt:
Subject :
----- Message Text -----
set  fish-ecology  repro
```

Conferencing Exercise 3. IDENTIFY AND SUBSCRIBE TO A USENET NEWSGROUP

Task You are a physicist working on cold fusion experiments and want to discuss measurement techniques with colleagues. How might you find and get started on a discussion group for workers in this field?

Approach Search one of the directories of discussion groups to find something of possible interest. Then read its FAQ document (if available) to get a better idea of the topics discussed before subscribing (and posting) to the group.

Connect to the searchable version of the Dartmouth Siglist.

```
% lynx  http://alpha.acast.nova.edu/cgi-bin/lists
Making HTTP connection to alpha.acast.nova.edu.
Sending HTTP request.

               SEARCH LIST OF DISCUSSION GROUPS

Use this page to search for interesting e-mail lists to join.

Details
      * Enter a phrase to search for.
      * The search is case insensitive.

Source

The database of lists employed here comes from the list of
Bitnet and Internet Interest Groups maintained at Dartmouth
College. There are currently more than 5900 entries! The
database is updated weekly.
```

Press the '**s**' key to start a keyword search.

```
This is a searchable index.  Use 's' to search      s

Enter a database search string: fusion

http request sent; waiting for response


Data transfer complete                    Listservs (p1 of 6)
Mailing Lists Matching fusion
_____

Fusion - Redistribution of sci.physics.fusion
```

The directory listing indicates that the listserv group "fusion@vm1.nodak.edu" operates in parallel as the Usenet newsgroup "sci.physics.fusion".

```
FUSION

  Addresses
      * FUSION@VM1.NODAK.EDU (list)
      * LISTSERV@VM1.NODAK.EDU (listserv)
  .
```

You prefer to access the group via Usenet.

```
  .     <text deleted>
  .
```

To get any available FAQ document, check the Usenet FAQ archives at MIT.

If an FAQ document is available, it will be in a directory corresponding to the group name.

For sci.physics.fusion, there is an extensive FAQ document available in multiple parts. It would be quickest to get the files with a single **mget** command.

Toggling off the **prompt** command means the system will not prompt you after each file is transferred.

You may now display the FAQ files back on your own computer, but will probably wish to download them to his PC so you can edit or print them from a word processing program package.

```
% ftp   rtfm.mit.edu
Connected to rtfm.mit.edu.
Name (rtfm.mit.edu:clementg): anonymous
331 Guest login ok, send your complete e-mail address as
password.
Password: clementg@solix.fiu.edu
230 Guest login ok, access restrictions apply.
ftp> cd   /pub/usenet/sci.physics.fusion
250 CWD command successful.
ftp> dir
200 PORT command successful.
150 Opening ASCII mode data connection for /bin/ls.
total 359
-rw-rw-r--   6 root      3           5604 Oct 20 01:13
Conventional_Fusion_FAQ_Section_(Intro)
.
.    <display deleted>
.
-rw-rw-r--   6 root      3          14456 Oct 20 01:13
Conventional_Fusion_FAQ_Section_5_11_(Devices-Status)
-rw-rw-r--   6 root      3          13920 Oct 20 01:14
Conventional_Fusion_FAQ_Section_6_11_(Recent_Results)
-rw-rw-r--   6 root      3          13956 Nov  3 01:58
Conventional_Fusion_FAQ_Section_7_11_(Education)
-rw-rw-r--   6 root      3          12616 Nov  3 01:59
Conventional_Fusion_FAQ_Section_8_11_(Internet_Resources)
-rw-rw-r--   8 10746     3          18470 Dec 24 21:00
FAQ_New_for_alt.inventors _Dec._1994_(end_of_year!)
226 Transfer complete.
2336 bytes received in 0.66 seconds (3.5 Kbytes/s)
ftp> prompt
Interactive mode off.
ftp> mget   Conventional_Fusion_FAQ*
200 PORT command successful.
150 Opening ASCII mode data connection for
Conventional_Fusion_FAQ_Section_(Intro)
226 Transfer complete.
.
.    <display deleted>
.
ftp> quit
221 Goodbye.

% more   Conventional_Fusion_FAQ_Section_(Intro)
```

Once you have looked over the FAQ, you are ready to start following the discussions online with a newsreader such as **tin** or **rn**.

The **tin** Group Selection menu brings up the designated group. The '**u**' in the left column indicates that you are currently unsubscribed. To subscribe immediately, simply press '**s**'.

The **tin** screen organizes the postings into topical threads -- shown here are the first five threads, with their corresponding subject lines.

The next column indicates whether you have already read a thread (**+** means you haven't).

The next number indicates how many individual postings, or articles are available for that thread (including the original article plus any follow-up posts).

The last column, marked generically with **[author]** in this example, would show the author of the article.

Here is the first article in the thread '**MRA and AC measurements**'.

```
% tin  sci.physics.fusion
tin 1.2 PL2 [UNIX] (c) Copyright 1991-93 Iain Lea.
Connecting to newshost...
Reading news active file...
Reading attributes file...
Reading newsgroups file...


          Group Selection (newshost  1)    h=help

u  1 196 sci.physics.fusion  Info on fusion, esp."cold" fusion.


                 *** End of Groups ***



          sci.physics.fusion (84T 197A 0K 0H R)    h=help

1  +      Conventional Fusion FAQ Section 0/11 (Intro)  [Author]
2  + 2    Calorimetry offer                             [Author]
3  + 4    Borrowing a scope                             [Author]
4  + 6    MRA and AC measurements                       [Author]
5  +      Bringing tools                                [Author]
.
.     <display deleted>
.
```

```
Thu, 05 Jan 1995 17:17:02  sci.physics.fusion  Thread 5 of 84
Lines 16          MRA and AC measurements  5 Responses
jsmith@mindlink.bc.ca  Jean Smith at MIND LINK! - British
Columbia, Canada

Has anyone performed over-unity power measurements on AC
signals by putting the power supply for the experiment ...
.
. <text deleted>
.
```

To post an article in response, press the **f** key.

You are in the default editor set for your newsreader (in this case, it is the Unix editor **vi**).

The text of the original message appears first, set off with colons (**:**).

You may edit some of the original message and/or add your own response.

When you issue the normal exit command for the editor, the newsreader asks whether you wish to send the posting. To do so, press '**p**'.

```
--
Jean Smith jsmith@mindlink.bc.ca
-------------------------------------------------------------
<n>=set current to n, TAB=next unread, /=search pattern,
^K)ill/select,a)uthor search, B)ody search, c)atchup,
f)ollowup, K=mark read,|=pipe, m)ail, o=print, q)uit, r)eply
mail, s)ave, t)ag, w=post


f



Subject:  Re: MRA and AC measurments
Newgroups: sci.physics.fusion
References: <62315-789580608@mindlink.bc.ca>
Distribution: world
Jean Smith (jsmith@mindlink.bc.ca> wrote:

:Has anyone performed over-unity power measurements on AC
:signals by putting the power supply for the experiment ...

[Add  text  of  response  here]

~

~

~

Z Z


Your article will be posted to the following newsgroup:
sci.physics.fusion
q)uit,  e)dit,  p)ost: p
```

Conferencing Exercise 4. START A NEW USENET NEWSGROUP

Task Your group of engineers wishes to start a new Usenet Newsgroup in the 'sci.' hierarchy, and you don't know where to start.

Approach Consulting some kind of "Frequently-Asked Questions" (FAQ) document is usually a good strategy when you don't know where to start. Without knowing exactly what newsgroup might produce a FAQ addressing your query, however, you may be reluctant to go into the Usenet FAQ FTP archive at MIT and start digging through the many directories organized by newsgroup. Your best bet is to go to an FAQ archive site that permits browsing and keyword searching, as demonstrated below.

The FAQ archive that has been formatted for the World-Wide Web will probably provide full-text descriptions of the FAQ's, available for browsing or searching.

```
% lynx   http://www.cis.ohio-state.edu/
                        hypertext/faq/usenet/

                    USENET FAQS

This document contains a list of all USENET FAQs found in
news.answers. The document is alphabetized by topic (more or
less).

Many of the FAQs in this list are presented in the same format
as they appear in the newsgroup, while others have been further
processed and split into additional documents. For more
information on all aspects of this project, see the technical
notes.

A few of the documents are provided in hypertext by the FAQ
maintainers, rather than in converted plaintext. Those
documents are shown with titles in italics.

Please send comments and complaints to fine@cis.ohio-state.edu.

New!
```

Initiate a keyword search with the '/' key in Lynx, or by clicking on the ? icon in a graphical Web client.

Keep the search simple, and truncate if possible.

```
Enter search string: creat
```

Lynx jumps to a matching item -- just what your group was hoping for.

```
                    List of USENET FAQs (p11 of 38)

        *   Creating  Newsgroups
        * Inflammatory Bowel Disease FAQ V1.2
        * Crossword FAQ
        * Cryonics FAQ
        * Cryptography FAQ
        * Atari ST FAQ
        * FAQ: CSH Coke Machine Information
```

You can browse the document online and make sure it provides the information you need.

NOTE:To create a group in the 'alt' or **local news** hierarchies, one does not have to go through this process.

In those cases, see your local systems administrator to set up the newsgroup. It will not receive the global distribution of the conventional Usenet hierarchies, but for course-related or other local groups, the limited distribution might be appropriate.

Your group does wish to pursue establishment of a new group in the sci. hiearchy, and
you need to save this document for a closer read. Press '**p**' to call up the menu for document retrieval options (the term 'printing options ' is a bit of a misnomer here).

Creating Newsgroups

    * Usenet Newsgroup Creation Companion
    *** How to Create a New Usenet Newsgroup**

        How to Create a New Usenet Newsgroup (p1 of 9)

HOW TO CREATE A NEW USENET NEWSGROUP

Archive-name: creating-newsgroups/part1
Original-author: woods@ncar.ucar.edu (Greg Woods)
Comment: enhanced & edited until 5/93 by spaf@cs.purdue.edu
(Gene Spafford)
Last-change: 17 Feb 1994 by tale@uunet.uu.net

        GUIDELINES FOR USENET GROUP CREATION

REQUIREMENTS FOR GROUP CREATION:

    These are guidelines that have been generally agreed upon across USENET as appropriate for following in the creating of new newsgroups in the "standard" USENET newsgroup hierarchy. They are NOT intended as guidelines for setting USENET policy other than group creations, and they are not intended to apply to "alternate" or local news hierarchies. The part of the namespace affected is comp, news, sci, misc, soc, talk, rec, which are the most widely-distributed areas of the USENET hierarchy.
.

.   <text deleted>

.

Arrow keys: Up and Down to move. Right to follow a link; Left to go back.
 H)elp O)ptions P)rint G)o M)ain screen Q)uit /=search [delete]=history list **p**

           PRINTING OPTIONS

    There are 161 lines, or approximately 3 pages, to print. You have the following print choices:

Save to a local file

Mail the file to yourself

Print to the screen

Please enter a file name: **creategroup.faq**

Conferencing Exercise 5. JOIN A LIVE CONVERSATION

Task See what conversations are taking place 'in real time' on environmental issues.

Approach Log on to the EnviroLink Network's EnviroChat; invoke online help if needed. Since this is a 'live' site with an ever-changing roster of visitors and conversation topics, your experience will be unique and unpredictable.

Connect to EnviroChat by telnetting to the appropriate address.

Log in with your first name and the password of your choice.

The other inhabitants logged on to EnviroChat will see you, and will speak to you with the name you entered upon logging on.

Note that whatever you say is echoed by the system. Also, your conversation may be continually interrupted with various announcements concerning other visitors' activities.

To get started, try any of the following commands:

.news summarizes purpose, scope, and policies for EnviroChat.

Notice that the purpose and rules of the group are clearly stated, to help provide some quality control.

```
% telnet envirolink.org  2000


Give me a name: gail
Give me a password:


Shanespeare walks in
Shanespeare says: hello gail
hello
You say: hello
Sign off: Thom is dyxs, dis...can't spell.
Shanespeare asks: whatcha up to?
well, I was hoping for a little tutorial or some basic
introduction to this chat service.
You say: well, I was hoping for a little tutorial or some basic
introduction to this chat service.
Shanespeare says: you shpold go to the campfire
Shanespeare says: here's for environment
Sign on (meadow): Trish the dancer
Okay, thanks for your help.
You say: Okay, thanks for your help.

.news

*** News ***
Use .help to see all the commands.
Use .help general to get a short explanation of commmands.
Use .help <command> to see specific help on the specified
<command>


******************
This Chat host is meant for ENVIRONMENTAL discussion only.

** There are administrators who monitor ALL traffic on this
system and you will be kicked off of the system if you violate
any of our policies.  The actions of administrators are beyond
argument and are considered to be the final word.

These are our policies:
- Environmentally-related topics only
- Show respect for others and their beliefs (no
```

.help provides a quick summary table of available commands.

.help general provides brief descriptions of each command.

.map provides the layout of the Chat environment -- very useful for those needing visual orientation.

Leave whenever you please.

---

harassment,etc.)
   - NO ON-LINE SEX will be tolerated

.help

*** Commands available ***

```
quit        who        shout       tell        listen
ignore      look       go          private     public
invite      emote      rooms       knock       write
read        topic      search      review      help
.
.       <display deleted>
.
```

All commands start with a '.' and can be abbreviated
For further help type .help <command> or .help general for general help.

.help general
GENERAL HELP:

All commands start with . or /   Commands and user's names can be abbreviated.
Use .help <command> for more detailed usage of the specified command.

Speech commands:

```
.tell    Send a private message to someone on the talker.
         Usage: .tell <person> <message>   You can substitute >
         for .tell
.emote   Display a posed message to those in your current area.
         Usage: .emote <action>  You can substitute ; or : for
         .emote
.
.       <display deleted>
```

.map
```
                              council          ___ amphitheater
Use .go me to
                                |             /
return directly
                    farm       |    mists    /
to the meadow
                       :..      \      \     /
                        :......\___ meadow ....... briars
            river ____                  |  \
                       \  swamp        /    \  cliffs    treehouse
                        \  |          /      \    \        /
                         \  |        /        \    \      /
                                            _ ___  campfire
```

.quit

Conferencing Exercise 6. PARTICIPATE IN A VIRTUAL MEETING

Task Visit a virtual conference center to converse with colleagues in real-time. The sample session shown here is one small illustration of the people, objects and activities you might encounter.

Connect to BioMOO via Telnet (for alternate access to the MOO, point your WWW browser to `http://bioinfo.weizmann.ac.il:8888`).

Read the introductory menu for ideas on how to get started.

```
%telnet  bioinfo.weizmann.ac.il  8888

Welcome to...BIOMOO [graphic]
          ...the virtual meeting place for biologists.

Type:
'purpose'         to read a statement on BioMOO's purpose
'connect (userid) (password)' to connect, for example:
                             connect Elmer stu888ph
'connect Guest'  to connect as a guest,
'create'          for information on how to get your own userid,
'who'             just to see who's logged in right now,
'@quit'           to disconnect, either now or later.

For human help please email to
Gustavo@bioinformatics.weizmann.ac.il.

purpose
----------------------------------------------------------------
BioMOO is a virtual meeting place for biologists, connected to
the Globewide Network Academy. The main physical part of the
BioMOO is located at the BioInformatics Unit of the Weizmann
Institute of Science, Israel.

BioMOO is a professional community of Biology researchers.  It
is a place to come meet colleagues in Biology studies and
related fields and brainstorm, to hold colloquia and
conferences, to explore the serious side of this new medium.
.

.      <display deleted>

.

Welcome!
```

Begin a session as a guest.

```
connect  Guest
*** Connected ***

The Lounge
A large, silent, dimly illuminated room with lots of people
snoring their real  lives away... A door to the south leads out
to the Central Room. (type 'south' to exit the Lounge)
A big sign here reads: Type 'tutorial' to learn the basics of
MOO.
You see a Guest Book and a list of registrants with bad email
address here.
```

You may choose a name, for use by other MOO dwellers

To get started, you might read through the tutorial.

Or, for a quicker introduction, check the online help.

Once you get familiar with basic commands, you can start roaming.

Each room contains objects for viewing, reading, or using in some way.

For an overview of the MOO, why not take a tour on the MOObus (it s free, but not always available!).

Hop on board.

```
                 Notice: To provide your name, type: @rename
                         Otherwise, you stay in anonymity.

@rename
[Type your name or `@abort' to abort the command.]
Gail
You set your name to Gail.

help  introduction
BioMOO is a 'text-based virtual reality' for biologists, in
which the users can move about from place to place manipulating
their environment and communicating with other users.

BioMOO is not a game, but a serious application of a technology
that has been largely devoted to gaming. People here meet to
cooperate, learn, teach, explore, create.

The first five kinds of commands you'll want to know are listed
below.  Type  'help <topic-name>' for details on any of them:

look -- getting a description of the current room or any object
say -- speaking to the other people in the same room as you
who -- showing which users are currently connected to the MOO
movement -- how to move around in the MOO, from room to room
@quit -- disconnecting from the MOO

south
You move south.

The Central Room
A very large, circular room, its ceiling a transparent dome
through which  sunlight streams in. A large archway leads west
into a foyer. There are doors leading in all directions,
labelled with tasteful signs.
You see the Research Directory, a public bulletin board, a
Jobs/Postdocs  bulletin board, a Lost and Found Box, a road-
sign pointing to the special interests room here, and the
MOObus here.

A sign on the MOObus catches your eye...
 "To take a short tour of the MOO type 'enter moobus' and
follow instructions."

enter  moobus
The MOObus
A relatively new bus. It's painted in a schoolbus orange.
Although it has red/orange fire streaks down each side.
You enter the MOObus.

Operating manual for MOObus
'tour'        shows available tour routes you can take
'show_tours'  shows you the itineraries for each tour
.
.        <display deleted>
```

Choose a tour.

```
tour
Pick a tour from the following list:
2.  Tour 2 (20 mins - 10 stops) by Martin
3.  Tour 3 (18 mins - 9 stops) by Martin
4.  Tour 4 (14 mins - 7 stops) by Martin
.
.       <display deleted>
.
[Type a line of input or `@abort' to abort the command.]4

The tourguide says: ''Welcome to MOObus Tours. The tour will
begin in 30 secs.
You'll see some highlights of the moo.''

You will have 2 minutes at each location. You may de-bus by
'out' and re-board with 'enter bus'. Make sure to re-board
within the 2 minutes to continue your tour.'

MOObus moves. Outside MOObus, you see:

Epstein-Barr Virus (aka EBV)

The tourguide says: 'You have arrived in #385. You may 'exit`
but be sure to 'enter bus` within 2 minutes!'
exit

The Epstein-Barr Virus (aka EBV)
You have plunged through a lipid bilayer and are now within the
viral envelope.
You see an Introduction, a latent mRNA, and the MOObus here.
DavidH (sleeping) is here.
You exit from the MOObus.

read  Introduction
There appears to be some writing on the note ...

The Epstein-Barr virus is a gammaherpesvirus with a double-
stranded DNA genome 172 kb in size. As with other
herpesviruses, the genome is packaged within a protein capsid
enveloped by a lipid membrane. EBV encodes roughly a hundred
.
.       <display deleted>
.
(You finish reading.)

Outside: The tourguide yells, ''Your tour will resume in about
30 seconds.  Type 'enter bus' to resume the tour. If you miss
the bus, type '@go #1715' to  resume your tour.''
enter  bus

MOObus moves.
Outside MOObus, you see:
```

Hop off the bus to look around.

There are various objects to read and look at in this room.

Uh-oh! Got a bus to catch!

Get off the bus to visit this lab.

Note that you get a full description of the lab and all that is in it.

You have many options for movement, for looking at or reading objects, etc.

The owner of this lab is in.

Maybe he could answer a few questions....

Better make it quick -- the bus is about to leave.

Enter the lab owner.

You must preface your text with 'say'. The echoed line shows what the other MOO dwellers see.

Note that this text was interrupted by the message that the bus was leaving.

If you feel sheepish for interrupting a lab owner, head directly home to read the tutorial.

When you are ready to leave the MOO, type @q.

```
Lab (WCL)
A bright and cheery place with a large desk and a comfy chair.
.
.       <display deleted>
.
```
exit
```
WatsonCrick's Lab (WCL)

A bright and cheery place with a large desk and a comfy chair.
On the desk resides a Mac, a stained coffee cup decorated with
the structure of a caffeine molecule, a MOO manual and a slew
of space-filling molecules.  Beside the desk is a spiral
staircase leading up to the MUSE room. [type: up or u to
enter]. To the WEST is the entrance to The Chemistry Laboratory
[type: w or west to  enter].
You see a Huge Blackboard, a Poem About Life, and the MOObus
here.
WatsonCrick (working - please page) is here.


        Welcome to the lab guest
        You are welcome anytime
```
page WatsonCrick
```
Your message has been sent to WatsonCrick (idle 5m, working -
please page).

The tourguide yells, ''Your tour will resume in about 30
seconds. Type 'enter bus' to resume the tour. If you miss the
bus, type '@go #1715' to resume your tour.''

WatsonCrick says, "what can I help you with? :0"
```
Say I am just trying to get MOObus drives off into the
distance **a sense of what scientists use a MOO such as
this for.**
```
The guest [Gail]: I am just trying to get a sense of what
 scientists use a MOO such as this for.
WatsonCrick says, "attend seminars, form journal clubs and make
models"
```
**say, Hmmm, sounds useful... how do you think most
newc**WatsonCrick says, "R U a biochemist?"
**omers start out. Is there a 'mentor' to'show them
around'?**
```
The guest [Gail]: Hmmm, sounds useful... how do you think most
newcomers start out. Is there a 'mentor' to'show them around'?
WatsonCrick says, "there's a tutorial in the lounge that is the
best way to learn a few basics"
```
**say, yes, that's a good place to start.Sorry for the
interruption. Thanks again.**
```
The guest [Gail]: yes, that's a good place to start.Sorry for
the interruption. Thanks again.
```
home
```
You click your heels three times.
```
tutorial

Module 2: Finding Colleagues

IN THIS MODULE:

Introduction
Strategies
Tools
Finding tools
Exercises

Introduction

At present, the most prevalent means of communication on the Internet is electronic mail. Each day, several million Internet users in every continent exchange email messages with colleagues, business associates, friends, and family worldwide.

What is most problematic about using electronic mail on the Internet is finding the exact address of the person one wants to contact. There is no centralized, all-inclusive white pages of Internet users (just as there is no global directory of telephone phone numbers or street addresses). As a result, searchers need to apply various tools, in trial-and-error fashion, to track down a colleague's email address.

 6

Strategies

The most effective strategies for finding colleagues on the Internet require some information about the persons being sought—their full or last names, where they work, and so on. This approach plays well in the favor of scientists, whose names and affiliations feature prominently on conference badges, on publications, or in the 'author' and 'corporate source' fields of many scientific indexing and abstract services. However, until those sources also include email addresses, sci/tech professionals will need to use one of the strategies described below.

The following questions may provide some guidance in determining which one(s) to try, or in explaining confusing, contradictory, or erroneous search results.

What is the scope of the finding tool or directory service?

How completely does it cover its designated group of users?

How much data is available (and searchable) for each entry?

How updated and accurate is the data?

How easy is the search interface to use?

Does the service report on the results when a search is unsuccessful?

 7

There is no ideal method for conducting a no-stones-unturned search for someone on the Internet. Typically, a search requires a combination of tools:

- Electronic mail to query a listserver
- Telnet connection to a distributed directory service (either directly or through a gateway on Gopher or the World-Wide Web)
- Gopher-based indexing tools such as the CSO Directory
- WAIS-based directories available directly, or through a Gopher gateway.

As more personal home pages and other directory information becomes available in hypertext format from HTTP servers on the World-Wide Web, Web indexing tools such as WebCrawler and the WWW Worm may also become effective for finding people on the Internet.

Each search system has distinct advantages and limitations. Some may require only a partial name, but perform the search in just one directory. Others may search directories that are incomplete, inaccurate, or out-of-date. Yet other search systems may cover a larger number of directories, but require precise information upfront to limit the possible number of hits. Before trying a particular directory service, it is therefore advisable to become acquainted with the scope of the search and the minimum criteria needed to perform it.

 8

TOOLS

Organizational Phonebooks & Directories

Searching an organization's employee directory is often the most effective way of finding someone on the Internet. Many colleges, universities, government agencies, and other research organizations have made their phonebooks publicly available on a network server.

One of the most common types of 'electronic phonebook' are the CSO directories, used in conjunction with Gopher systems and easily recognized by the <CSO> suffix on the Gopher menu. Developed by the

Computer Services Office at the University of Illinois, CSO server software is now in wide use by institutions worldwide to provide a keyword-searchable directory of their employees, staff, or students. Each CSO directory works more or less the same, offering common searchable fields such as personal name, email username, department, and more. However, not all institutions include data for all fields, so some CSO-based phonebooks vary in the amount of information they provide about their constituents.

Other types of keyword-searchable phonebooks may also be in use on the Internet. On Gopher, these may be denoted with a <?> suffix on the menu. Each directory may operate somewhat differently, offering more or fewer fields, or providing differing types of data about the individuals included in the database. Simple searches on first and last name may prove most successful.

As official directories for a particular institution, organizational phonebooks tend to be complete, accurate, and regularly maintained and updated. But because they are limited in scope to the employees of a particular institution, they are helpful only if you already know where your colleague works.

Professional Directories

If you know your colleague's profession, try looking in a directory for that area of specialty. Professional directories may be either formally maintained by a particular association or society, or informally maintained by volunteers. Examples of directories available for science and technology professions include the American Mathematical Society's combined membership list

```
gopher://e-math.ams.org//77/profInfo/amscml
```

and the directory of biologists' addresses, linked to Harvard's Biosciences subject hub

```
http://golgi.harvard.edu/biopages.html
```

Like other electronic phonebooks, professional directories may take a variety of forms, from non-searchable text files to keyword-searchable databases with varying fields for searching. Simple searches on first and last name are recommended.

Professional directories tend to be incomplete. Inclusion in the directory may be voluntary or self-initiated, omitting anyone who didn't ask or agree to be included. Also, unless formally maintained by an organization or group, professional directories may not systematically purge old data, leading to inaccurate search results.

Discussion Group Subscriber Lists and Archives

If you think the person you are seeking follows a particular discussion group, there is a good chance you can retrieve the email address from the group's subscriber list. For listserv discussions groups, one may obtain a list of all subscribers (except those who have set their subscriptions to conceal their names) by sending a simple email message to the listserv managing the group. For Usenet groups, request from the

```
mail-server@rtfm.mit.edu
```

the email address of anyone who has posted a message to any Usenet newsgroup within the last year.

Though this technique is 'hit-or-miss,' it may be effective for an individual without a permanent or long-term affiliation to an institution or profession, such as a graduate student or someone new to the profession.

Distributed Internet Directory Services

Distributed Internet directory services provide a single search interface to discrete, heterogenous directories available at different sites across the Internet. They range in their development from experimental to fully operational. Among the most popular of these services are Netfind, Fred, Whois++, Knowbot, and Paradise. You can find descriptions of these services in a number of guides available on the Internet.[1]

None of the distributed directory services on the Internet performs a global search. Each uses a variety of search protocols and algorithms that are often hidden from the user. When a search is unsuccessful or returns erroneous data, the reason is often difficult to understand or explain. The best approach may be to try several of the directory services in succession. If one search fails to turn up the desired address, another may prove successful.

Finding Finding Tools

Thanks to the development of several 'hubs' for finding people on the Internet, searching for colleagues by 'trial-and-error' has been made a lot more efficient and easy. Numerous links to organizational phonebooks, gateways to distributed directory services, and even some email how-to guides can be found on the menus of these sites:

Washington and Lee's "Searching for PEOPLE" site

```
gopher://liberty.uc.wlu.edu/11/internet/
personsearches
```

Texas Tech's "Phone Books/Directories" menu

`gopher://cs4sun.cs.ttu.edu/11/Phone%20Books`

Notre Dame's "Phone Books" Collection

`gopher://gopher.nd.edu/11/`

`Non-Notre%Dame%20Information%20Sources/`

`Phone%20Books—Other%20Institutions`

Professional directories and discussion group archives may be a little harder to track down, because they may be embedded somewhere in a Gopher menu hierarchy or on a World-Wide Web page. For guidance in finding these, refer to Section C of this guide.

Endnote

1. The following guides available on the Internet provide clear, to-the-point descriptions of the various directory services on the Internet:

   December, John, "Network Information Retrieval Utilities," Internet Tools Summary (updated regularly).

   `http://www.rpi.edu/Internet/Guides/decemj/itools/nir.html`

   EARN Staff, "Guide to Network Resource Tools," Internet Engineering Task Force (IETF), Network Working Group, FYI 23, RFC 1580 (March 1994).

   `gopher://ds.internic.net/00/fyi/fyi23.txt`

   Foster, J., editor, "A Status Report on Networked Information Retrieval: Tools and Groups." Internet Engineering Task Force (IETF), Network Working Group, FYI 25, RFC 1689 (August 1994).

   `gopher://ds.internic.net/00/fyi/fyi25.txt`

Finding Colleagues Exercise 1. FIND AND USE AN INSTITUTIONAL PHONEBOOK

Task Is there a colleague you would like to contact? See if you can track down the email address from his or her institution's phonebook.

Approach To see if your colleague's phonebook is available on a Gopher, look for it at one of the "Phonebook" hubs. Note whether it is a CSO Directory or another type of Gopher directory service. To see how this works, follow along with the example shown here, in which you are searching for a fictional colleague by the name 'Avamarie Smith' at Boston University.

Point Gopher to the Phonebooks 'hub' at the University of Notre Dame.

```
% gopher gopher.nd.edu
--> Non-Notre Dame Information Sources/
      --> Phone Books--Other Institutions/
            --> North America/
```

Note that Boston University uses a CSO directory.

```
      Beth Israel Hospital (Harvard Univ.) <CSO>
      Board of Governors Universities (Illinois) <CSO>
--> Boston University <CSO>
      Bradley University <CSO>
      Brigham Young University <CSO>
      Brown University <CSO>
      Bucknell University <CSO>
      Bull HN Information Systems <?>
      California Institute of Technology <?>
```

CSO directories may offer many fields for searching or display, but not all are available at every institution.

Since you don't know how your colleague's first name might be entered in the directory, keep the search simple the first time around.

```
+------------------Boston University-------------+
|
|   name                    smith
|   academic_rank
|   title
|   department
|   office_address
|   office_phone
|   deptid
|   college
|   reg_status
|   RESTRICT
|   local_address
|   local_phone
|   home_address
|   alias
|   email
|   other_unix_aliases
|   index_id
```

The directory returns a list of possible matches to browse through. As noted here, you must select the desired entry and key in the index-id in the CSO search box to retrieve the email address. Other CSO directories may not require this step.

The abbreviated entry provides the index_id.

In the CSO search box, you now enter only the individual's index_id.

Now the full entry displays. Compare with the CSO search box displayed above. Note that data is not actually available for all of the fields shown.

```
Boston University (35k)
+---------------------------------------------------------+
E-mail addresses and telephone numbers are only
displayed when a query matches one individual.  To
obtain additional information on a particular
individual, inquire again with the index_id.

 Query help for additional suggestions.

 -----------------------------------------------------
                name: Smith, Avamarie
       academic_rank: Assistant Professor
          department: Chemistry
            index_id: X9999


   .
   .     <display deleted>
   .

  | index_id              X9999

  | type
 -----------------------------------------------------+


Boston University
+---------------------------------------------------------+

               name: Smith, Avamarie
              title: Assistant Professor
         department: Chemistry
     office_address: 999 Commonwealth Ave
       office_phone: (617) 353-9999
              alias: amsmith
              email: amsmith@bu.edu
           index_id: X9999
```

Finding Colleagues Exercise 2. USE WAIS TO SEARCH FOR A PROFESSIONAL DIRECTORY

Task Determine if any professional directories in your field are searchable by WAIS.

Approach Search the WAIS Directory of Servers to identify any WAIS source that contains the word 'addresses'. Then search that source for a particular address.

To see how this works, follow along with this example, in which you are searching for the address of a marine biologist named Smith who works at Woods Hole Oceanographic Institution. (Note: All information displayed here is fictional! Feel free to substitute your own search).

Connect to the public WAIS client at WAIS, Inc.	`% telnet wais.com`
	
	`login: wais`
	
	`Welcome to swais, the text-terminal telnet client to WAIS.`

Telnet connections usually require login id, password, and terminal emulation	`Please type user identifier (optional, i.e user@host):`
	`clementg@solix.fiu.edu`
	`TERM = (vt100) vt100`

Note that this client is running simple WAIS, or swais.	

```
                Source Selection                    Sources: 524

    #          Server           Source              Cost
  001:  [          archie.au]   aarnet-resource-guide   Free
  002:  [ndadsb.gsfc.nasa.gov]  AAS_jobs                Free
  003:  [ndadsb.gsfc.nasa.gov]  AAS_meeting             Free
  004:  [    munin.ub2.lu.se]   academic_email_conf     Free
  .
  .       <display deleted>
  .
```

The client loads all 500+ WAIS sources.

If a source is already known, jump right to it by pressing the '/' key.

At the prompt, enter the name of the source. WAIS jumps right to it.

```
Source Name: directory-of-servers

  194:  [        irit.irit.fr]  directory-irit-fr      Free
  195: * [    quake.think.com]  directory-of-servers   Free
```

Select the desired source with the **<spacebar>**; your selection will be marked with an asterisk. Then press 'w' to enter your search term(s).

```
Keywords: addresses
```

This is a broad, preliminary search to identify any WAIS source that indexes people's addresses.

```
Search Results                        Items: 23
  #     Score   Source             Title            Lines
001:  [1000]  (directory-of-se)  linux-addresses      43
002:  [ 945]  (directory-of-se)  biologists-addresses 16
003:  [ 556]  (directory-of-se)  astropersons         58
004:  [ 444]  (directory-of-se)  internet-mail       131
  .
  .       <display deleted>
```

WAIS returns matching entries from the Directory

of Servers.

If you see a directory of interest, move to the desired line and press **<return>** to view description of the source from the Directory of Servers.

The description provides technical information about where the WAIS-searchable database actually resides, and descriptive information about the source for your reference.

Press the **<spacebar>** to select this source for searching. Then enter as many specific terms as possible because 'woods' is a common name, and may match irrelevant entries.

The top-ranked item appears to be the address we wanted. Its score of 1000 indicates that WAIS deems this directory entry to be the most relevant of those retrieved, based on the number of occurrences of the search terms.

Selecting line 001 retrieves the desired listing. It may be saved by pressing '**S**' or sent by electronic mail by pressing '**m**'.

```
Getting "biologists-addresses.src" from directory-of-servers.src

(:source      :version  3
  :ip-address "134.172.2.69"
  :ip-name "net.bio.net"
  :tcp-port 210
  :database-name "biologists-addresses"
  :cost 0.00
  :cost-unit :free
  :maintainer "biosci@net.bio.net"
  :description "Server created with WAIS release 8 b5 on Jul 31
  23:21:10 1993 by shibumi@net.bio.net.  The files of type
para used in the index were:
  /var/spool/uucppublic/pub/BIOSCI/biosci-user.addresses.
  This is an address directory of biologists who use the
BIOSCI/bionet newsgroups dedicated to research in
biology/biological sciences.")


Keywords:  smith  woods  hole  oceanographic  whoi

SWAIS                Search Results                Items: 15

#    Score      Source          Title                 Lines

001:   [1000] (biologists-addr)  NAME: Smith -  Maribel    26
002:   [ 773] (biologists-addr)  NAME: Smith  - Frank -  Q 26
003:   [ 728] (biologists-addr)  NAME: Woods  - Edward -  P 26
.
.      <display deleted>
.

Getting "NAME: Smith - Maribel from biologists-addresses.src...

NAME:  Smith - Maribel
date (DD-MM-YY): 11-15-94
first name: Maribel
middle initial:
family name: Smith
job title: Research Associate
e-mail address: msmith@cliff.whoi.edu
e-mail network: Internet
phone number: 508-548-1400, ext. 9999
FAX number: 508-457-9999
institution: Woods Hole Oceanographic Institution
```

Finding Colleagues Exercise 3. FIND A SOCIETY'S PROFESSIONAL DIRECTORY

Task See if there is a professional directory available from one of the scientific societies representing your field.

Approach Check the *Scholarly Societies* hub at the University of Waterloo to see if there is a link to the society of your choice, and then check to see if its membership directory is available online.

If you can't find a directory for your field, follow along with this example for the American Mathematical Society.

Telnet to the publicly-accessible server at the University of Waterloo.

(For alternate access methods, see the note below)

```
%telnet uwinfo.uwaterloo.ca (or 129.97.128.100)
login: uwinfo

-->Electronic Resources Around the World/
  -->Campus and other information systems../
    -->Gophers of Scholarly Societies/
      -->American Mathematical Society../
        -->Professional Information../
          -->AMS Combined Membership List <?>
```

Since you don't know what fields are available in the directory database, it is best to keep the search simple.

```
Search for:  smith



  -->  Matches for people with last name smith .

Penelope  Smith
Department of Mathematics
University of Puget Sound
Tacoma, WA 98416-0001
206-756-3562
audreys@ups.edu
```

The search results are returned in a file available for browsing online, or saving to your own computer account.

```
Timothy Smith
Department of Mathematics
University of Keele
Keele ST5 5BG
England
t.smith@keele.ac.uk
.
.       <display deleted>
.


% gopher  uwinfo.uwaterloo.ca
% lynx  http://www.lib.uwaterloo.ca/society/
                             overview.html
```

Alternate Access Methods:

Finding Colleagues Exercise 4. SEARCH A USENET SUBSCRIBER LIST FOR A
NEW COLLEAGUE'S EMAIL ADDRESS

Task You want to locate the email address of a young colleague you met recently at a conference, but don't know where she works.

Approach Your young colleague may follow a USENET Newsgroup. Send a search request to the mail-server at MIT, where an archive of USENET postings is maintained.

From whatever mail program you use, send a search request to the mailserver at MIT.

Leave the subject line blank, and in the message field enter the search term.

```
%mail  mail-server@rtfm.mit.edu
Subject:
send  usenet-addresses/galileo
```

A mail message will soon arrive with a somewhat ragged listing of USENET subscribers, with personal name, email address, and date of the last article posted to a USENET newsgroup.

Note that matches to the name 'Galileo' may occur in any part of the subscriber's name.

```
galileo@compengg.Eng.Sun.COM (Galileo Mills)(Apr 4 94)
galil001@DMI.USherb.CA (LUIGI GALILEO) (Apr 9 94)
dgs5@cornell.edu (David Galileo Smith)  (Apr 9 94)
ghsu@waterloo.hp.com (Galileo Hsu)      (Apr 24 94)
gg6900@swuts.sbc.com (Genny A. Galileo) (Apr 23 94)
wongg@eecg.toronto.edu (Galileo Wong)   (Apr 14 94)
```

Finding Colleagues Exercise 5. SEARCH A LISTSERV SUBSCRIBER LIST FOR AN EMAIL ADDRESS

Task   You are looking for the e-mail address of a colleague whose work history you have lost track of, but who is apt to subscribe to a particular Listserv Discussion List.

Approach   Request the list of subscribers from the Listserv where the discussion group is maintained. Try this exercise with a Listserv group you know, or identify one from the 'Directory of Scholarly Conferences', available at:

```
gopher://gopher.usask.ca/11/Computing/Internet20%Information/
Directory%20of%20Scholarly%20Electronic20%Conferences
```

Or follow along in the example shown here: search for a nuclear engineer on the discussion group, ANURT-L@VM1.HQADMIN.DOE.GOV, the "Forum on Advanced Nuclear Reactor Technology".

Send email message to the listserv where the group is maintained. Leave the subject line blank, and type the review command in the message field. If in UNIX mail, press **control-d** to send the message.

```
%mail  listserv@vm1.hqadmin.doe.gov
Subject:
review  anurt-l
```

A mail message will then be returned with a listing that looks like this.

```
Date:          Tue, 19 Dec 1994 08:47:33 -0400
Subject:       File: "ANURT-L LIST"

*  DOE Advanced NUclear Reactor Technology Discussion
*
*  Review= Public     Subscription= Open         Send= Private
*  Notify= Yes        Reply-to= List,Ignore      Files= No
*  Validate= Store
*  Confidential= Yes
*  Notebook= Yes,W,Monthly
*
*  Owner=  U7542NF@DOEVM (Norman Fletcher)
*
```

Note that the listing returned by the listserver includes the personal names and email addresses of all but the "concealed" subscribers.

```
jsmith@ACM.ORG                       John Smith
KITCHENSSM@A1.OSTI.GOV               Steve Kitchens
YOUNGENG@A1.OSTI.GOV                 Greg Youngen
QRS3%QA%DCPP@BANGATE.PGE.COM         Q. Smith
.
.         <display deleted>
.
*
```

It also provides the total number of subscribers on the list, as well as contact information about the listowner.

```
* Total number of "concealed" subscribers:      2
* Total number of users subscribed to the list: 32
(non-"concealed" only)
* Total number of local node users on the list: 2
(non-"concealed" only)
```

Finding Colleagues Exercise 6. USE A DISTRIBUTED DIRECTORY SERVICE TO FIND AN EMAIL ADDRESS

Task You want to find the email address for a colleague for whom you know the place of work.

Approach Try running a search using the distributed directory service Netfind, which requires some information about your colleague's name and location to perform a successful search.

Make an 'educated' guess at the name of the computer holding your colleague's email account. Remember that the last part of the domain name (the top level domain) reflects the type of organization (e.g., 'edu' for U.S. academia; 'gov' for U.S. federal research labs; 'mil' for U.S. military personnel; 'org' for U.S. non-profit research institutions; 'uk' for sites in the United Kingdom, etc.). For the rest of the domain name, try the institution's first name (e.g., 'stanford' or 'weizmann') or the common acronym (e.g., 'ornl' for Oak Ridge National Laboratory, etc.)

Or follow along with the example below, which demonstrates a successful search for fictional colleague Isaac Miller, a Professor of Geology at the Weizmann Institute of Science in Israel.

Connect to one of the many Netfind servers available.

```
%telnet  bruno.cs.colorado.edu  (or  128.138.243.150)
login: netfind
```

Choose the option to 'search'.

```
             1. Help
    -->      2. Search
             3. Seed database lookup
             4. Options
             5. Quit (exit server)
```

Enter keywords for your colleague's name and location.

```
Enter person and keys (blank to exit) --> miller weizmann il
```

Netfind displays a list of possible domains for the computer holding your colleague's account. Using cues from the listing, select the most promising domain(s) to search.

In this example, you avoided domains 2, 3 and 4 because these servers appear to belong to departments other than Geology.

```
Please select at most 3 of the following domains to search:
0. ubique.co.il (ubique limited, weizmann institute campus,
rehovot, israel)
1. weizmann.ac.il (the weizmann institute of science, israel)
2. cs.weizmann.ac.il (computer science department, the weizmann
institute of science, israel)
3. math.weizmann.ac.il (mathematics department, the weizmann
institute of science, israel)
4. weizmann.weizmann.ac.il (computing center, weizmann
institute of science, rehovot, israel)
5. wisdom.weizmann.ac.il (the weizmann institute of science,
israel)
```

Domains 1 and 5 appear to for general campus use.

Netfind uses a variety of search protocols and search algorithms to check all computers within the selected domains, and reports back if it finds a matching username.

In this case, a query using Finger returned a few matches.

Netfind found a successful match to the search in the domain wisdom.weismann.ac.il.

Netfind provides all information available from Finger, including the user's login name and address, real name, location, time of last login, and any information the user has included in his or her plan.

Based on the time and date of last login, and the note included in Miller's plan, the correct email address for this colleague is: miller@wisdom. weizmann.ac.il

The next two matches Netfind returns are for another 'Miller'. The first account is not in use -- the uowner has never logged in.

The second account is active, judging by the time and date of last login, but the owner has not included any text in his plan.

```
Enter selection (e.g., 2 0 1) --> 1 5


( 2) SMTP_Finger_Search: checking domain wisdom.weizmann.ac.il
( 1) got nameserver wisipc.weizmann.ac.il
( 1) got nameserver weizmann.weizmann.ac.il
( 1) got nameserver nufar.wisdom.weizmann.ac.il
( 1) SMTP_Finger_Search: checking domain weizmann.ac.il
( 3) SMTP_Finger_Search: checking nameserver
wisipc.weizmann.ac.il
( 4) SMTP_Finger_Search: checking nameserver
weizmann.weizmann.ac.il
Mail is forwarded to shapiro@WEIZMANN.weizmann.ac.il
( 4) SMTP_Finger_Search: checking host WEIZMANN.weizmann.ac.il


SYSTEM: wisdom.weizmann.ac.il
Login name: miller          In real life: Miller Isaac
Office: 211Z, x4998         Home phone:  x3131
Directory: /wisdom/users3/miller      Shell: /bin/tcsh
Last login at Wed Feb 8 13:14 from
miller@at1.wisdom.weizmann.ac.il
Plan:
      Isaac Miller, Professor of Geology
      Office: Weizmann, Ziskind bldg.-14, phone (08)-342221
      Home phone: (08)-473106
      (This is my newest address. The previous one is
invalid).
```

```
Login name: gold            In real life: Miller Gold
Directory: /silver/users/gold         Shell: /bin/tcsh
Never logged in.
No Plan.

Login name: ajm             In real life: Miller Abraham
Directory: /silver/users/ajm    Shell: /bin/tcsh
Last login at Wed Feb 22 16:38 from
ajm@silver.wisdom.weizmann.ac.il
No Plan.
```

Netfind also reports the domains for which it can not perform a search.

Finally, Netfind provides a report summarizing the results of the search.

Netfind is ready to try again.

```
( 4) do_connect: Finger service not available on host
WEIZMANN.weizmann.ac.il ->
 cannot do user lookup


FINGER SUMMARY:
- Found multiple matches for "miller" in finger output, so
unable to determine most recent/last login information or most
promising electronic mail information.  Please look at the
above finger search history and decide for yourself which is
best.

Continue the search ([n]/y) ? --> n
```

Module 3: News and Current Information

IN THIS MODULE:

Introduction
What's available
Access methods
Quality and value
How to find
Exercises

Introduction

The category of news and current information includes informal communication and publication sources that serve to announce, update, or provide the latest directions or instructions. Though not directly used in the research process, these resources play an important role in the daily life of a sci/tech professional, facilitating planning, problem solving, and other day-to-day activities

Traditionally, scientific news and current information has taken the form of printed trade publications, newsletters, bulletins, and mailings, issued from scientific or related agencies, organizations and publishers. As these information-issuing organizations set up shop on the Internet, users may now gain instant access to much of this information from their desktops, often in advance of its appearance in printed or other sources. The Internet has thus become a primary source of new information for the scientific community.

The Internet is also enabling smaller organizations and groups to become global publishers of news and current information. Academic departments, informal professional networks, even individuals, may now get their message out quickly and efficiently using a variety of Internet tools. New information on the Internet is therefore reaching a wider audience than ever possible with traditional mass media.

What's Available

News and current information found on the Internet cover a wide range: immediate descriptions of observable phenomena, such as volcanic eruptions or comet collisions; important research developments, such as a successful laboratory experiment; announcements concerning conferences or calls for papers; opportunities for funding, Requests for Proposals, employment listings, or fellowships; latest versions of proposal forms; updates to government policies or regulations; press releases and product announcements; and more.

Access Methods

News and current information on the Internet may take a variety of formats and access methods, depending on the provider of the information and how they choose to establish a presence on the Internet. The time-sensitive, wide-interest nature of new information requires a quick, far-reaching, and easy-to-update means of delivery.

'Broadcasting' information on the Internet most commonly takes the form of discussion groups and email distribution lists, where a single news posting may be quickly multiplied and delivered to hundreds or thousands of users worldwide.

Users then automatically receive a steady stream of news and current information which they may filter for use.

Announcements and updates may also be 'posted' to an organization's Internet server, and made available using Finger, FTP, Gopher or WWW. This requires users to visit the site to browse any new information that has become available.

A third method of disseminating new information on the Internet are the emerging tools for real-time digital broadcasting: Internet Relay Chat, MOO, and Multicasting (MBONE). Already the latter is used to broadcast space shuttle launches and radio talk shows. As more users gain access to and familiarity with these tools, live broadcasting on the Internet promises to become an important source of news and current information for the scientific community.

Quality and Value

Separating the wheat from the chaff is always a concern on the vast, unregulated Internet, and some of the greatest variations in quality occur with the quick-to-produce, informal sources such as news and current information.

The most important qualities of any news source are timeliness and reliability, balanced equally to ensure the value and usefulness of the information provided. Whereas news reporters in the traditional media are obligated to verify new information from at least one source, no such mechanism is in place on the Internet. It is therefore incumbent upon any user of news on the Internet to recognize that available information may have varying degrees of quality, and therefore develop some practical criteria for gauging reliability.

Most importantly, one should always "consider the source." Publishers, societies and other agencies well known in the scientific community are likely to produce high-quality news and current information. In the unlikely event that the information is inaccurate or incomplete, at least there is a formal, permanent organization that can respond to a question or request for more information.

Information from smaller information providers, groups, and individuals may be equally as reliable, but there is no guarantee. Users should heed any disclaimers that appear on the menu, in a README file, or on the hypertext pages of such services; they should also note the contact address for the site administrator or maintainer in case later clarification is needed.

Ultimately, each user of news and current information on the Internet has to determine the usefulness of the available sources, weighing the trade-off between timeliness and reliability to determine if the information meets his or her needs.

How to Find News and Current Information

Finding very recent information on the Internet may be difficult because of the very fact that it is so newly available. Search tools such as Veronica or a World-Wide Web spider may miss a very recent item because their indexes, though updated quite frequently (weekly in Veronica's case), are not created on demand at the time of the search. Also, if the new information was delivered via a discussion group, one would have to anticipate which group(s) received the posting and determine where and how to search those archives.

The most successful strategy for finding new information on the Internet may be to identify the possible organizations disseminating the information, and then locate their sites online. This step may not be difficult for sci/tech users, who are generally well-acquainted with the names and geographic locations of the important information publishers in their fields (*e.g.*, scientific societies, funding agencies, federal research programs or laboratories, publishers).

Locating the network presence of such known entities is usually successful with some trial and error. Some of the best places to start are:

- Scholarly Societies hub at University of Waterloo

  `gopher://watserv2.uwaterloo.ca:70/00/servers/`
  `campus/scholars`

  `http://www.lib.uwaterloo.ca/society/overview.html`

- Government Gophers section on Gopher Jewels

  `gopher://cwis.usc.edu:70/11/Other_Gophers_and_`
  `Information_Resources/Gopher-Jewels/government`

- Grant section on Gopher Jewels

  `gopher://cwis.usc.edu:70/11/Other_Gophers_and_`
  `Information_Resources/Gopher-Jewels/research/grants`

- Listing of "Gophers Around the World", arranged geographically

  `gopher://gopher.tc.umn.edu:70/11/Other%20Gopher`
  `%20and%20Information%20Servers`

- Listing of all WWW servers, arranged geographically

  `http://info.cern.ch/hypertext/DataSources/WWW/`
  `Servers.html`

Many more Internet finding strategies are described in Section C of this guide.

News Exercise 1. SEARCH USENET GROUP ARCHIVES FOR CURRENT
INFORMATION

Task   A colleague just asked if you were planning to present a paper at the Information
Systems conference at Johns Hopkins this summer. You hadn't even seen the Call
for Papers! Now you want to find out the details of this conference.

Approach   Search the archives of the USENET Newsgroup ieee.announce for the Call For
Papers. The archives are likely to be maintained on the IEEE Gopher.

Connect to the IEEE gopher.	`% gopher  gopher.ieee.org`

Move through the menus until reaching the desired document, 'CALL FOR PAPERS' (Note that the period '.' at the end of the menu line denotes a file)	`--> ListProcessor and Newsgroup ARChives/`
	`    --> Newsgroup ARChives: ieee.announce/`
	`        --> ieee-announce.9406/`
	`            --> CALL FOR PAPERS - CISS 1995.`
	
	`From: eed_wbh@jhunix.hcf.jhu.edu (Brian Hughes)`
	`Newsgroups: ieee.announce`
	`Subject: CALL FOR PAPERS - CISS 1995`
	`Date: 10 Jun 1994 22:59:40 -0400`

Gopher allows you to browse the full text of the document.	

```
              CALL FOR PAPERS - CISS '95

Conference on Information Sciences and Systems
The Johns Hopkins University
Department of Electrical and Computer Engineering
Baltimore, Maryland
March 22, 23, 24, 1995

Authors are invited to submit papers describing new advances,
applications, and ideas in the fields of communications,
computer engineering, signal and image processing, and systems
and controls.

Two kinds of papers are solicited.
(1) Regular papers requiring approximately thirty minutes for
presentation; these will be reproduced in full (up to 6 pages)
in the Conference Proceedings.
(2) Short papers suitable for presentation in approximately
fifteen minutes; one-page summaries of these papers will be
.
.                      <text deleted>
.
```

Use Gopher's **<s>** option to save the document as a file on your own computer system.

```
Press <RETURN> to continue, <m> to mail, <s> to save, or <p> to
print:s
```

You may name the saved file however you wish.

```
Save in File: cisscall.txt
```

News Exercise 2. SEARCH THE GOPHER OF A RESEARCH LABORATORY

Task You submitted (unsuccessfully) a grant proposal for developing a thin film lithium battery, and one of the reviewers remarked that you hadn't acknowledged similar work already achieved at Oak Ridge National Laboratory. How might you track down information about this research project?

Approach Since you know the laboratory where the work was performed, you can go to its Internet site (in this case, the Oak Ridge Gopher) to get descriptions of research completed and in progress.

Connect to the Oak Ridge National Lab's Gopher	`% gopher  gopher.ornl.gov`
	
	`--> Catalog of Oak Ridge National Laboratory Technologies`

Select the desired directory	`    --> Complete Text Search <?>`

Perform a key word search (Gopher runs a simple WAIS, or 'SWAIS' search here)	`Words to search for:   `**`lithium batter`**`y`
	
	`  Complete Text Search: lithium battery`

Gopher retrieves all files matching your keywords	`  --> 1.   THIN FILM RECHARGEABLE LITHIUM BATTERY.`
	`      2.   HIGH-TEMPERATURE THERMAL STORAGE PERFORMANCE TESTING.`
	
	`THIN FILM RECHARGEABLE LITHIUM BATTERY`
	
	`ORNL has developed rechargeable thin film lithium batteries that have potentially many applications as power sources for electronic devices.  The batteries, which are typically less than 10 microns`
	`.`
	`.      <display deleted>`
	`.`

Use Gopher's **<s>** feature to save the document to a file on your own computer system.	`Press <RETURN> to continue, <m> to mail, <s> to save, or <p> to print:`**`s`**
You may name the saved file as you wish.	`Save in file: `**`battery.txt`**

News Exercise 3. FIND VERY RECENT NEWS BY TRACKING DOWN THE
GOPHER OF THE NEWS-ISSUING ORGANIZATION

Task Try finding some very recent news on the Internet.

Approach In this example, you are looking for a recent news piece about the
proposed abolition of the United States Geological Survey in the Republican
Party's 'Contract With America'. You tried Veronica first, searching all of
Gopherspace to pick up any discussion group postings that might have been
archived on Gopher.

When that was unsuccessful, you tried finding the site for an organization
likely to have a press release or other information about this news story.

Connect to Veronica
and enter a search.

```
% gopher  gopher.unr.edu
--> Search ALL of Gopherspace (4800 servers) using Veronica/
```

Tried different search
strings with no success.
This is just one example.

```
Search Words:

aboli*  geological  survey
```

This item may be too recent
to be picked up in
Veronica's last indexing
run.

```
*** Your search on "aboli* and geological survey" returned
nothing/
```

Now look for a Gopher site
for the American
Geological Institute, an
organization likely to issue
or collect press releases and
other new items of interest
to the geology profession.

```
% gopher  gopher.tc.umn.edu

--> Other Gopher and Information Servers/
    --> North America/
        --> USA/
            --> All/
```

Though you don't know the
geographic location of the
Institute, you can look for it
in the alphabetical list of
Gophers in the U.S.

To search for desired menu
item, press '**/**' and enter the
name. Gopher jumps right
to it.

```
Search for:

american  geological  institute


--> American Geological Institute/
```

Aha! The news item is featured prominently on the first menu.

```
--> Abolition of U.S.Geological Survey and U.S.Bureau of Mines


----------------------------------------------------------------+
                  (c) American Geological Institute


                   Congress Threatens to Abolish
         U.S. Geological Survey and U.S. Bureau of Mines

                      Craig M. Schiffries
                 American Geological Institute
```

This item may be saved with an 's' or sent to another user's email address with an 'm'.

```
The U.S. Geological Survey and the U.S. Bureau of Mines are
facing one of the most serious challenges in their history.
Both agencies have been targeted for complete elimination
according to an attachment to the Contract with America.  The
Contract contains a package of 10 bills that 224 Republican
members of Congress have pledged to introduce in the first 100
days of the new Congress.

.
.       <text deleted>
.


Mail current document to:
```

News Exercise 4. USE WAIS TO JOB-HUNT

Task You are a recent PhD, looking for a post-doctoral research position in Astrophysics, and you want to check the job register maintained by the American Astronomical Society (AAS).

Approach To perform a keyword search over full-text documents on the Internet, WAIS is a good tool to use. However, since you do not know the exact name of the AAS source, you will have to browse through the list of all sources.

Connect to the public WAIS server located at UNC.

For alternate access methods, see the note below

This WAIS site loaded 663 possible sources to search! Fortunately, the AAS should appear pretty close to the top of the alphabetical list.

Using the down arrow key, move down to the desired source (AAS-jobs), and press the `<spacebar>` to select. Pressing the `<return>` key loads the database, and pressing the 'w' key then displays the prompt for a keyword search.

The system returns a number of matches, with the item ranked most relevant at the top.

Look at the position at Princeton -- move down to it and then select it by pressing the `<return>` key.

```
%telnet  sunsite.unc.edu
login: swais
TERM = (unknown) vt100


SWAIS           Source Selection              Sources: 663
   #          Server            Source              Cost

001:   [            archie.au]  aarnet-resource-guide   Free
002:  *[ndadsb.gsfc.nasa.gov]  AAS_jobs                Free
003:   [ndadsb.gsfc.nasa.gov]  AAS_meeting             Free
004:   [      munin.ub2.lu.se]  academic_email_conf     Free
005:   [         sv3.cnusc.fr]  acubase                 Free
  .
  .      <text deleted>
  .


Keywords: postdoc astrophysics

Enter keywords with spaces between them; <return> to search; ^C
to cancel <return>



SWAIS          Search Results            Items: 14

#     Score        Source     Title                  Lines

001:[1000]  (AAS_jobs)  Faculty Position Astrophys - YALE      22
002:[ 715]  (AAS_jobs)  Faculty Position Astronomy and Astro   47
  .
  .      <text deleted>
  .
011:   [ 178]  (AAS_jobs) Post-doctoral Research COLUMB        33
012:   [ 178]  (AAS_jobs) Assistant Professor - PRINCETON      33
013:   [ 178]  (AAS_jobs) Post-doctoral Fellowship STOCKHOLM   24
```

A full-text description of the job listing then appears.

Note that this is NOT a post-doctoral position. It was retrieved because it includes a match to the search term 'astrophysics', and SWAIS connects multiple search terms with an OR. The low relevance score of this item (Score=178) is an indicator that this is not what was wanted.

A search with just the term 'postdoctoral' yields better results.

Alernate Access Method:
Gopher & Jughead

```
SWAIS                              Document Display
Page: 1
----------------------------------------------------------------
Assistant Professor - PRINCETON UNIVERSITY
No. 8337
Assistant Professor
PRINCETON UNIVERSITY
Department of Astrophysical Sciences
Peyton Hall
Princeton, NJ 08544-1001

Princeton University anticipates appointment of a tenure-track
Assistant Professor in the Department of Astrophysical
Sciences, to begin in September 1994.   The primary criterion
for selection will be a demonstrated capability for original
astronomical research as an optical-infrared observer and
promise of future activity and growth.
.
.        <text deleted>
.
send letters of recommendation. Consideration of applications
will begin in mid-February 1994, and all material should arrive
by March 1, 1994, at the latest.


Keywords:  postdoctoral


SWAIS          Search Results                    Items: 14

#      Score Source       Title                             Lines

001:[1000] (AAS_jobs)    Post-doctoral Appointment - UNIVER   28
002:[1000] (AAS_jobs)    Post-doctoral Research  - COLUMB     33
012:[ 780] (AAS_jobs)    Post-doctoral Fellow - JOHNS HOPKINS 33
.
.        <text deleted>
.


%  gopher  sunsite.unc.edu
 -->   Search the Gopher menus using Jughead <?>

Words to search for

      AAS

 -->  1.  AAS_jobs.src <?>
      2.  AAS_meeting.src <?>

Words to search for

      postdoctoral
```

News Exercise 5. USE FTP (OR GOPHER) TO RETRIEVE AN NSF FORM.

Task You are preparing a grant proposal to submit to the National Science Foundation, and need a clean copy of the form 'GPG Information About PI's'.

Approach Thanks to the NSF's information system (STIS) on the Internet, you can get a postscript version of the form immediately. If you already know the exact name of the form (it's 1225), you can use FTP to get a quick copy. Then you just need to open the file in your wordprocessor, enter the appropriate data, and send it to a postscript printer to get a clean final copy.

Connect to NSF's FTP server and login as anonymous, using email address for a password.

```
% ftp stis.nsf.gov
STIS.NSF.GOV>anonymous
<Guest login ok, send your E-Mail address as password.
Password: clementg@servax.fiu.edu
< Guest login ok, access restrictions apply.
<Welcome to the STIS System
Logged into x.nsf.gov.
NcFTP 1.8.2 (August 4, 1994) by Mike Gleason, NCEMRSoft.
x.nsf.gov:/NSF/forms
```

Note that the server automatically puts you in the forms directory.

```
ncftp>dir
total 1352
-rw-rw-r--  1 101      50        241 Nov 10 13:56 fm1030
-rw-rw-r--  1 101      50      50497 Nov 10 13:56 fm1030.ps
-rw-rw-r--  1 101      50        228 Jun  3 16:44 fm1207
-rw-rw-r--  1 101      50      48516 Apr 22  1994 fm1207.ps
-rw-rw-r--  1 101      50        185 Apr 22  1994 fm1207s
-rw-rw-r--  1 101      50      49595 Apr 22  1994 fm1207s.ps
-rw-rw-r--  1 101      50        218 Apr 22  1994 fm1225
-rw-rw-r--  1 101      50      32554 Apr 22  1994 fm1225.ps
-rw-rw-r--  1 101      50        186 Apr 22  1994 fm1225s
-rw-rw-r--  1 101      50      32580 Apr 22  1994 fm1225s.ps
-rw-rw-r--  1 101      50        216 Apr 22  1994 fm1239
-rw-rw-r--  1 101      50      33102 Apr 22  1994 fm1239.ps
-rw-rw-r--  1 101      50        717 Apr 22  1994 fm1239s
-rw-rw-r--  1 101      50      33964 Apr 22  1994 fm1239s.ps
.
.      <display deleted>
.
```

For each form listed, a text and postscript version are available.

Get the smartform version of Form 1225, denoted with the 's' after the form number. This version includes dummy data filled in at the appropriate spot -- typing your data over it will ensure you fill out the form correctly.

Note that you use the default transfer mode, ASCII, with a .ps file.

```
ncftp>get fm1225s.ps
Receiving file: fm1225s.ps
100%  032580 bytes. ETA:  0:00
fm1225s.ps: 32580 bytes received in 7.53 seconds, 4.23 K/s.
x.nsf.gov:/NSF/forms
ncftp>quit
<Goodbye.
```

Now you may download the file to a PC, if necessary, open it with a text editor or word processor, and replace the dummy data with your own

Alternate Access Method: GOPHER

If you did not already know the exact name of the file needed, you could also have used the menu-driven gopher interface at the same location to browse the names of the forms.

Note that because the form is in Postscript format, it is not viewable on the Gopher screen.

It can, however, be readily saved as a file by positioning the cursor at the desired line and pressing 's' to save.

Or, if you have set-up your PC telecommunications software for downloading from your mainframe, you may use the Gopher download feature.

Instead of saving the file, press 'D' to download.

In the box displaying download options, choose the desired method -- this depends on what methods are supported by your mainframe and by the communications software on your PC.

Alternate Access Method: World-Wide Web

```
% gopher  stis.nsf.gov
--> NSF Publications/
     --> Forms

     1.fm1030- FM 1030 - GPG Summary Proposal Budget Form.
     2.fm1030.ps- FM 1030 - GPG Summary Proposal Budget Form.
     3.fm1030s- Form 1030.
     4.fm1030s.ps- Form 1030.
     5.fm1207- NSF Form 1207 - Cover Sheet for Proposals.
     6.fm1207.ps- NSF Form 1207 - Cover Sheet for Proposals
     7.fm1207s- Form 1207.
     8.fm1207s.ps- Form 1207.
     9.fm1225- FM 1225 - GPG Information about PI's/ PD's
     10.fm1225.ps- FM 1225 - GPG Information about PI's/ PD's.
     11.fm1225s- Form 1225s.
--> 12.fm1225s.ps- Form 1225s.
     13.fm1239- FM 1239 -  GPG Current and Pending Support.
     14.fm1239.ps- FM 1239 -  GPG Current and Pending Support
     15.fm1239s- form 1229.
     16.fm1239s.ps- form 1229.
     17.fm1263- FM 1263 -  NSF Grant Transfer Request Form.
     18.fm1263.ps- FM 1263 -  NSF Grant Transfer Request Form.

              Save in File: fm1263.ps


          Internet Gopher Information Client v2.0.16

                          Forms

1.fm1030- FM 1030 - (Revised)GPG Summary Proposal Budget Form
2.fm1030.ps- FM 1030 -(Revised)GPG Summary Proposal Budget Form
3.fm1207       +-------fm1225s.ps  - Form 1225s--------+osals
4.fm1207.ps    |                                       |osals
5.fm1207s      | -->    1.  Zmodem                      |
6.fm1207s.p    |        2.  Ymodem                      |
7.fm1225       |        3.  Xmodem-1K                   |s / PD's
8.fm1225.ps    |        4.  Xmodem-CRC                  |s / PD's
9.fm1225s      |        5.  Kermit                      |
10.fm1225s.p   |        6.  Text                        |
11.fm1239      |                                       |upport
12.fm1239.ps   | Choose a download method (1-6):       |upport
13.fm1239s     | [Help: ?]  [Cancel: ^G]               |
14.fm1239s.p   +---------------------------------------+
15.fm1263      - FM 1263 -  NSF Grant Transfer Request Form
16.fm1263.ps   - FM 1263 -  NSF Grant Transfer Request Form


Download Complete. 34060 total bytes, 34060 bytes/sec
Press <RETURN> to continue

% lynx    http://www.nsf.gov/
```

News Exercise 6. USING WWW, FIND CURRENT INFORMATION ABOUT A REMOTE FIELD AREA BEFORE YOU DEPART.

Task You are planning a research expedition to the McMurdo Sound area of Antarctica, and want some idea about their summer weather conditions.

Approach Without knowing who might publish such information (there are not apt to be many producers of news on the Internet in Antarctica), you are not sure where to begin. It might be best to start broad, with the manually-compiled WWW Catalog -- it includes many unique research-related sites.

Point your WWW browser to the Home Page for the W3 Catalog.

If you are using a text-only browser that can t handle forms, page down until you see the option for the **alternative search interface**. Select this option and then type **'s'** to begin a search.

Keep the search broad and simple, so you do not miss any possible sites inadvertently.

This looks like a good prospect.

Lynx does not display the clickable map available to users with graphical Web browsers.

```
% lynx   http://cuiwww.unige.ch/w3catalog
```

Enter a database search string: **Antarctica**

Result of search for "antarctica":

    January 25,1994: **Malin Space Science Systems**, developer of the camera on the ill-fated Mars Observer spacecraft, now has a WWW server. Among other things, the server contains MOC images taken during cruise, and an extensive set of stills and MPEG .

.

.   <display deleted>

.

    New Zealand: **International Centre for Antarctic Information and Research (ICAIR)**

The server contains annual programme reports for several nations outlining current projects and activities on the ice, information on Ozone depletion over Antarctica, and links to other Antarctic related Internet services. (**cvl**)

          Gateway to Antarctica - Home Page (p1 of 2)

    [IMAGE]

Click on the section of the map that is of interest, or for those without a graphical interface, select one of the following -

The **logistics** category looks custom-made for visiting researchers.

```
  * News
   * Education
   * Science
   * Treaty
   *  Logistics
   * Environment
```

LOGISTICS

* The on-line Antarctic Address Book.

* Current international exchange rates.

* Helicopters New Zealand.

* Floor plans of Scott Base on Ross Island.

> + Kitchen, dining, and recreation areas (35Kb).
> + Accomodation block (41Kb).

* **Daily weather reports** from Scott Base in the McMurdo Sound area
(77050'S, 166045'E).

Daily Scott Base Weather

Daily Scott Base Weather Reports

_____

Bingo! Here is a collection of weather reports.

These weather reports are posted by **NZAP** staff resident at Scott Base.

```
  * Nov, 17 Nov, ?997
  * Thu, 22 Dec, 1994
  * Wed, 21 Dec, 1994

  .
  .    <display deleted>
```

Message

Scott Base weather for the 22/12/94 (0900 Hrs)

_____

See what an early Summer day is like there....

Cheer up...maybe your next field area will be in Tahiti!

```
Air pressure: 986.0hp and steady.
Air temperature: -3.50C
Wind direction: 1500, speed: 0 Kts
Visibility: About 40 km
Cloud: 8/8
```

Module 4: Scientific and Technical Reference Tools

IN THIS MODULE:

Introduction
What's available
Access methods
Quality and value
How to find Internet reference sources
Exercises:

Introduction

This module focuses on factual, rather than bibliographic, information sources. The rich diversity of the latter, and their increasing integration with document delivery services, is treated separately in Module 5, "Searching and Retrieving the Literature."

What's Available

Reference sources are intended to be consulted for quick and ready facts and bibliographic information, rather than read from cover to cover. Traditionally, reference sources have taken the form of printed handbooks, encyclopedias, looseleaf services, as well as electronic databases, that typically remained in the library's Reference Department. Now many of these same sources are also available on the Internet from one's own desktop.

The Internet medium is, in fact, well suited for reference application, offering the advantages of quicker and more convenient access from one's office or laboratory. Some network reference sources offer greatly enhanced functionality and features such as keyword searching, hypertext links, and multimedia descriptions. And others represent entirely new tools, developed specifically to take advantage of the network's capabilities.

14

Access Methods

Because reference works are generally formatted to provide quick and easy access to specific information, certain Internet tools are especially well-suited for reference application. One of the most effective and popular reference tools is Gopher, whose hierarchical structure provides a natural progression from the general to the more specific. Formatting each section of a reference work as a separate Gopher menu item (*e.g.,* each specimen in a catalog, or each sheet in a collection of Material Data Sheets) makes it easier for users to see what they need, and also makes the information accessible with a keyword search using the Gopher search tool, Veronica.

Hypertext-formatted reference sources on the World-Wide Web are also effective, offering logical links to related information and 'collections' of materials on a similar topic. The keyword-search capabilities of most Web browsers enable users to skip to a relevant section of a Web page, and the numerous indexing tools for WWW enable searchers to quickly locate sites of interest.

For the briefest reference information, Finger offers a quick and easy means of access. Summary tables of current data, such as the year's hurricane forecasts or recent earthquakes, are commonly made available with Finger.

WAIS is a moderately effective tool for reference application. Its capabilities for full-text searching and relevance feedback may prove successful in tracking down very specific or hard to find information. And for reference information provided in image or other non-textual format (such as the Smithsonian's gem catalog or a collection of astronomical images), WAIS may be the only search tool that allows searches of corresponding text descriptions. In most cases, however, WAIS is not an effective reference tool because of its limited (and in some WAIS clients, lack of) Boolean capabilities and other search features. A search with simple WAIS, for example, is likely to retrieve many irrelevant documents that a user must view before discarding.

Access methods that offer a file only in its entirety—mailservers and FTP, for example—are useful when you need to quickly retrieve a cover-to-cover document. For providing quick reference information, they are less effective because they allow minimal, if any, browsing, and no online searching capabilities. However, you may apply a few techniques to enhance the reference use of a document once it has been retrieved to your own computer system. Using system commands such as *grep* in Unix, or the search commands in an editing or word processing program, you may perform simple keyword searches on the document saving you the trouble of reading through it to find desired information.

15

Quality and Value

To gauge the quality and value of sci/tech reference sources on the Internet, ask yourself the following questions, adapted from Katz[1] and Starr[2]:

1. Does the work fulfill its purpose, as stated or implied in its title, opening screens, first menu item, or README file?

2. Are the authors and/or publishers fully acknowledged somewhere online, and do they have sufficient knowledge and reputation to fulfill the purpose of the work?

3. What does the work offer over other sources, either on the Internet or in other formats?

4. For what audience is the source designed, and what is their assumed level of subject- or Internet-expertise?

5. What is the cost of the source, whether in direct user fees and/or in software/hardware/telecommunications expenses related to access over the Internet?

6. Does the network format offer ease-of-use, effective search or browse capabilities, and convenient options for retrieval and saving?

How to Find Internet Reference Sources

Because of their specialized nature, most sci/tech reference sources tend to be included on the menus of subject hubs or other subject-oriented sites. The Yale Chemistry Gopher,[3] for example, offers a menu of "Factual and Other Data," with links to an electronic periodical table of the elements, material data sheets and other fact sheets, chemical safety information and more.

Additionally, there a number of libraries on the Internet that serve as "Virtual Reference Desks," offering links and pointers to diverse reference sources from a single menu. For a single Reference hub that provides pointers to many "Virtual Reference Desks" under one roof, connect to the Solinet Gopher:

```
gopher://sol1.solinet.net/11/On-Line%20Ready%20Reference
```

Reference sources may also be tracked down using many Internet search tools. See Section C, "Finding Internet Resources in Your Area of Interest" for more guidance.

Endnotes

1. Katz, William A. *Introduction to Reference Work*, 6th edition.
    New York: McGraw-Hill, 1992.

2. Starr, Susan S., "Evaluating Physical Science Reference Sources on the Internet."
    The Reference Librarian, No. 41/42 (1994): 261-273.

3. The Yale Chemistry Gopher is available from URL

    ```
    gopher://yaleinfo.yale.edu:7700/11/OtherYaleGophers/chem/
    Reference/Data
    ```

Reference Exercise 1. USE FINGER TO FIND A QUICK FACT

Task What kind of hurricane season are we in for this year?

Approach Let's check the forecasts from Gray's research group at Colorado State.

```
% finger  forecast@typhoon.atmos.colostate.edu
```

```
Login name: forecast                 In real life: Forecast status
Directory: /users/forecast
Never logged in.
No unread mail
Plan:
************************************************************************
```

STATUS OF GRAY'S ATLANTIC SEASONAL HURRICANE FORECAST FOR 1994	1944–1993 Mean	Nov 24 1993 Fcst.	Jun 7 1994 Fcst.	Aug 5 1994 Fcst.	Observed
Named Storms	9.3	10	9	7	1
Named Storm Days	46.1	60	35	30	2.0
Hurricanes	5.7	6	5	4	0
Hurricane Days	23.0	25	15	12	0
Major Hurricanes (Category 3-4-5)	2.2	2	1	1	0
Major Hurricane Days	4.5	7	1	1	0
Hurricane Destruction Potential	68.1	85	40	35	0

```
************************************************************************
```

Reference Exercise 2. RETRIEVE A TEXT DOCUMENT AND USE THE UNIX GREP COMMAND TO SEARCH IT FOR A DESIRED KEYWORD.

Task As an environmental consultant, you are preparing an environmental impact statement, and need to find out the endangered status of the key deer.

You know you can get a current copy of the U.S. Fish & Wildlife Service's "List of Endangered and Threatened Wildlife and Plants" available from the agency's mailserver at r9irmlib@mail.fws.gov. Once retrieved to your own computer, you may then search it by keyword using the UNIX grep command.

Approach Because you don't know what commands the mailserver uses for retrieving or searching documents, you use 'trial-and-error' to request a list of instructions from the mailserver. You first try sending an email message to the mailserver with the word 'help' in the message field. This is the common procedure for requesting help from many listservers. If the mailserver at the Fish & Wildlife Service does not recognize that command, it may send, in response, a message indicating what commands it <u>does</u> use.

From the mail system on your own system (in this case, the program PINE), send a message to the mailserver at the Fish & Wildlife Service.

```
% pine

   PINE 3.90    MAIN MENU    Folder: INBOX  0 Messages


?  HELP              -  Get help using Pine
C  COMPOSE MESSAGE   -  Compose and send a message
I  FOLDER INDEX      -  View messages in current folder
L  FOLDER LIST       -  Select a folder to view
A  ADDRESS BOOK      -  Update address book
S  SETUP             -  Configure or update Pine
Q  QUIT              -  Exit the Pine program

Copyright 1989-1994.  PINE is a trademark of the University of
Washington.
```

Compose a new message.

As a first attempt, type the message '**help**' in the message field, leaving the subject line blank.

```
c

   PINE 3.90    COMPOSE MESSAGE   Folder: INBOX   0 Messages

To        : r9irmlib@mail.fws.gov
Cc        :
Attchmnt:
Subject :
----- Message Text -----
help


^G Get Help  ^X Send      ^R Read File ^Y Prev Pg   ^K
```

Later, a response from the mailserver arrives by email, indicating that it did not understand the original request.

The message does, however, describe the procedure for requesting help from this mailserver.

Prepare a new email message to send to the mailserver, with 'SEND HELP' in the subject line. (not shown here).

In response, you receive a lengthy set of user instructions and commands.

Instead of reading through the entire help file delivered as a mail message, you can search for a desired keyword in the message. The PINE mail program allows you to perform such a search.

But if your mail program does not have a function for searching messages, you will have to export or extract the message into a text file in your directory (consult your mail program manual for specific commands).

You may then search the file using the UNIX grep command (or, alternatively, the search capabilities of any available text editor).

```
PINE 3.90    MESSAGE TEXT   Folder: INBOX   Message 1 of 2 ALL

Date: Fri, 2 Mar 1995 03:20:25 -0400 (EDT)
From: "SMTP%"R9IRMLIB@mail.fws.gov"
To: clementg@solix.fiu.edu
Subject: Re: Help

SORRY....

I did not understand your request - please check your
documentation and try again. (HINT:  Check your spelling.)

If you need assistance, send a new message to this address with
the SUBJECT LINE "Send Help".

PLEASE ENTER ONLY ONE REQUEST PER MESSAGE.

If you are replying to a message sent to you from this
address, you will not receive a response other than
.
. <text deleted>
.

   PINE 3.90    MESSAGE TEXT   Folder: INBOX   Message 1 of 2 ALL

Date: Fri, 2 Mar 1995 05:13:19 -0400 (EDT)
From: "SMTP%"R9IRMLIB@mail.fws.gov"
To: clementg@solix.fiu.edu
Subject: Re: Send Help

        Subject: Library Server User's Guide
        Author:  R9IRMLIB at 9AR~PBA
          (INTERNET: R9IRMLIB@fws.gov)
        Date:    02/27/95
============================================================

NEWCOMERS

Welcome to the U.S. Fish and Wildlife Service (USFWS)
Information Resources Management (IRM) Library Server.  This
facility is accessible to anyone with the ability to exchange
electronic mail with the USFWS.

The Library Server is now operational 24 hours per day 7 days
per week.

All inquiries or requests should be addressed to R9IRMLIB.
Within the Department of Interior, that username may be
addressed at ~IFWS.

ATTENTION: INTERNET USERS - Please read instructions at the
bottom of this page regarding UUDECODE and FILE SIZE issues.
Also, please note addressing instructions.
```

Using grep, specify the '-i' flag for a case-insensitive search, and '-5' to display five lines preceding and following the matching word to provide context.

Two matches to the search term 'endangered' are displayed, each with a window of surrounding text.

Here are the instructions for requesting the list of Endangered and Threatened species.

Now compose a new mail message to the mailserver, with the message 'SEND ES INSTRUCTIONS' (not shown here).

You soon receive a response from the mailserver, indicating the proper command to use for this request ('SEND T&E LIST'). You may then compose a final message to the mailserver, as shown here.

When the T&E list arrives by email, export it to the file 'telist.txt' following the same steps shown above.

```
                         The Internet address for this username is
                                    R9IRMLIB@fws.gov

All requests must be entered in the SUBJECT LINE of your
message, not the MESSAGE TEXT.
.
.      <text deleted>
.


%grep  -i  -5  endangered  fwshelp.txt


MESSAGE TEXT.

================================================================
SUBSCRIBERS - WHAT'S NEW
----------------------------------------------------------------
      MISSION INFORMATION
      *   SEND ES INSTRUCTIONS - Updated Endangered Species
          Information 2/14/95
----------------------------------------------------------------
      TECHNICAL INFORMATION
      *   SEND CONFIGURATION - Model Configuration Guidelines for
          Computer Systems
          (2/7/95)
----------------------------------------------------------------
      (This contains a WordPerfect file - 30Kb)
      SEND PRESS RELEASES   - You will receive current Press
      Releases from the Fish and WildlifeService Office of
      Public Affairs.  Dates and file sizes will vary.
      * SEND ES INSTRUCTIONS  - You will receive instructions
      on retrieving information regarding the Endangered
      Species Act, Threatened and Endangered Species
      Listings, and Species Maps. LAST UPDATED 2/14/95
      SEND BIOLOGUES          - You will receive 14 fact sheets
      with biological information on various species...


PINE 3.90   COMPOSE MESSAGE          Folder: INBOX  0 Messages

To       : R9IRMLIB@MAIL.FWS.GOV
Cc       :
Attchmnt :
Subject  : SEND T&E LIST
----- Message Text -----
```

The file is large (over 200 kb) -- too large to browse easily.

Once again, use grep to search the text file. The following criteria are specified: case insensitive; display two lines before and after the matching term; and the text must exactly match the species name contained within the quotation marks.

Grep quickly displays the matching text, with two records preceding the following the matching entry. You now need to check the beginning of the file to determine what fields are delimited by each set of commas and quotation marks.

You can type out the first few pages of the file using the **more** command. When done, press '**q**' to quit.

```
% dir
total 2419
drwxr-xr-x  9 clementg ia4010ul    2048 Mar  7 04:41 ./
drwxr-xr-x495 root     wheel       8704 Mar  6 15:09 ../
.
.
.
-rw-------  1 clementg ia4010ul  202212 Mar  6 12:28 telist.txt


% grep -i -2 "Odocoileus virginianus clavium"
telist.txt

"Deer, Formosan sika","#Cervus nippon
taiouanus^","Taiwan","Entire","E","50","NA","NA"
"Deer, hog","#Axis (=Cervus) porcinus annamiticus^","Thailand,
Indochina","Entire","E","15","NA","NA"
"Deer, key","#Odocoileus virginianus clavium^","U.S.A.
(FL)","Entire","E","1","NA","NA"
"Deer, marsh","#Blastocerus dichotomus^","Argentina, Uruguay,
Paraguay, Bolivia, Brazil","Entire","E","3","NA","NA"
"Deer, McNeill's","#Cervus elaphus^","China (Sinkiang,
Tibet)","Entire","E","3","NA","NA"


% more telist.txt

 PINE 3.90   MESSAGE TEXT  Folder: INBOX  Message 1 of 2 ALL

Date: Fri, 10 Jun 1994 05:13:19 -0400 (EDT)
From: "Gail P. Clement" <clementg@solix>
To: clementg@servax.fiu.edu
Subject:ASCII T&E SPECIES LIST

Author:  DrewryG at 9AR~FWE1
Date:    2/14/95 8:23 AM


The four attached files describe the endangered and threatened
wildlife and plants listings, as published in the Federal
Register (50 CFR 17.11 and 17.12), and current through January
31, 1995.  These files are updated monthly.
Three of the files (ANIMLIST.ASC, PLNTLIST.ASC, DELIST.ASC) are
ascii comma delimited files for the animal, plant and delisted
species.  Descriptions of the fields in these files follows
below.
.
.     <display deleted>
.
Database Structure for the ANIMAL LIST File: ANIMLIST.ASC
NOTE: Fields delimited with COMMAS

Field  Field Name  Type       Width

   1 INVNAME      Character     70   (inverted common name
--More--(0%) q
```

Reference Exercise 3. LOCATE A KNOWN REFERENCE RESOURCE USING
A WWW ROBOT

Task You are a botanist, interested in finding an electronic version of the
Australian Plant Name Index somewhere on the Internet. You don't have a
clue as to where or in what format this reference resource might be available.

Approach A Web robot might be able to find a link to this resource regardless of the
type of server it is available on. The WebCrawler, in particular, is a good first
choice because it indexes the full-text of Web page, providing a lot of context
for each link it finds. Also, the WebCrawler search interface allows the
searcher to connect the multiple words of the search phrase with an "AND".

Point your Web browser (in
this case, the text-only
browser, Lynx) to the
WebCrawler Home Page.

Note that a link to a
'simple search page' is also
available for users whose
Web browser doesn't
support forms.

Using the up/down arrows
to jump from one hyperlink
to the next (or the mouse,
for graphical browsers),
move the cursor to the
search entry line and type
in your search term(s).

Be sure the AND connector
is selected with the
asterisk (*****), and the
number of results is set as
desired (the default is 25).

When you are ready to run
the search, move to the
hyperlink '**Search**' and
press the enter key.

Note that WebCrawler
found many 'hits', and
ranked the site for "Carol
Oakes' Reference Tools" as
being the most relevant (it

```
% lynx  http://webcrawler.cs.washington.edu/
             WebCrawler/WebQuery.html

             WebCrawler Searching (p1 of 2)

             [IMAGE] SEARCH THE WEB

To search the WebCrawler database, type in your search keywords
here.

Type as many relevant keywords as possible; it will help to
uniquely identify what you're looking for. NOTE: if this page
doesn't have a place to type your query, try using this simple
search page.

         australian plant name index_____
         Search             (*)AND words together
         Number of results to return:  [25_]

   .
   .     <display deleted>
   .


                    WebCrawler Search Results (p1 of 3)

                  WebCrawler Search Results

The WebCrawler is sponsored by DealerNet and Starwave. Please
see the sponsor page for more details.

The query "australian plant name index" found 34 documents and
   returned 25:

   1000 Carol Oakes' Reference Tools
   0785 Biodiversity
   0771 Biodiversity
```

has the most occurrences of the search terms).

Selecting that link takes you to the WWW Page where WebCrawler found a match to "Australian Plant Name Index".

Note that the new site is 19 pages -- too long to browse through looking for the desired link!

Rather, to save time, invoke the search option to quickly locate the link to Australian Plant Name Index. Press '/' in Lynx, or select the '?' or 'Find' option in your graphical Web browser).

The Web browser jumps right to the desired link.

Selecting on that link will take you directly to the site for the Australian Plant Name Index.

To be sure you can return to the site, add its link to your bookmark or hotlist. In Lynx, move the cursor to the desired line and press 'a', and then '1'. In a graphical Web browser, select the appropriate option from the menu.

```
0375 gopher://life.anu.edu.au/0/anubf.txt
0268 Bio Resources, Alphabetical Listing
0154 Forestry information resources on the Interne
      .
      .       <display deleted>
      .
```

                    Carol Oakes' Reference Tools (p1 of 19)

                        CAROL'S REFERENCE SOURCES
_____

Carol's Choices; a Hands On Demonstration of Web Resources by
Carol Oakes, Boise State University Library Reference
Department
_____

This homepage contains my personal selection of representative
types of Internet resources applicable to reference service
provided at the BSU Reference desk. The choices are not
inclusive. They serve the following purposes:

* they are informative of the kinds of information resources .
      .
      .   <display deleted>
      .

Enter a search string: **australian plant name index**

                    Carol Oakes' Reference Tools (p17 of 19)

         + **Australian Plant Name Index searchable**
         + Catalogue of Australian Mosses searchable
         + Common Names of Australian Plants searchable
         + Plant Names in Current Use searchable
         + USDA Plant Families and Genus Names searchable
         + Viruses--Organic
         + Viruses--Animal
      .
      .   <display deleted>
      .

Save D)ocument or L)ink to bookmark file or C)ancel? (d,l,c):**1**

Arrow keys: Up and Down to move. Right to follow a link; Left
to go back.

H)elp O)ptions P)rint G)o M)ain screen Q)uit /=search
[delete]=history list **Q**

Reference Exercise 4. USE VERONICA TO SEARCH GOPHERSPACE
FOR KEYWORDS

Task You are an industrial chemist, planning to use 4,4'-Thiodianiline for some laboratory procedures, and are concerned about exposing your technicians to any harmful effects. How can you quickly find out about the hazards of this substance and any necessary handling procedures?

Approach For a topic so narrow and specific, a simple Veronica search for keywords in filenames might turn up relevant information. Using Veronica to do a wide search across 'Gopherspace' will retrieve individual files, such as archived discussion group postings or other small bits of reference information available somewhere on Gopher.

Connect to Veronica from a Gopher server.

```
% gopher  gopher.unr.edu
--> Search ALL of GOPHERSPACE.../
      -->Search GopherSpace by Title word(s) .... <?>
```

Choose the 'Gopherspace' option for a wider, more exhaustive search.

In the search box, enter a keyword.

```
Words to search for

4,4'-Thiodianiline
```

Veronica finds one matching Gopher item.

To be sure of returning to this link, add it to your Gopher bookmark file by pressing 'a'.

To determine the source of the information, press '=' (the equal sign) and view the technical data for the link. This item comes from the EcoGopher at University of Virginia.

This looks like some kind of chemical data sheet. To find out exactly what source it is from, one may need to point Gopher to the host at Virginia and browse menu lines in the directory

```
Search GopherSpace by Title word(s) (via ...) 4,4'-Thiodian...

 -->  1.  4,4'-Thiodianiline




#
Type=0
Name=4,4'-Thiodianiline
Path=0/library/gen/toxics/4,4'-Thiodianiline
Host=ecosys.drdr.Virginia.EDU
.
.      <display deleted>
.


4,4'-Thiodianiline (14k)
+------------------------------------------------------------+
  Common Name:    4,4'-Thiodianiline
  CAS Number:     139-65-1
  DOT Number:     None
  Date:           May, 1989
  ------------------------------------------
```

above the file for 4,4'-Thiodianiline.

Point Gopher to EcoGopher, and select the directories according to the path provided in the link ('=') information displayed from the Veronica search.

Note that the technical names specified in the path above do not approximate the descriptive names used in the Gopher menus, so one may need to dig a little to find the right menus.

Now it is evident that the file found earlier is part of a collection of EPA Chemical Substance Factsheets.

Had this reference source been embedded in a file differently named, such as one large text file labelled "EPA Chemical Factsheets", Veronica would not have found the information on this substance. It was only because each factsheet is stored in a separate file, labelled with the name of the substance, that the Veronica search was successful.

```
HAZARD SUMMARY
  *    4,4'-Thiodianiline can affect you when breathed in and by
       passing through your skin.
  *    4,4'-Thiodianiline should be handled as a CARCINOGEN WITH
       EXTREME CAUTION.

.
.      <display deleted>
.


% gopher  ecosys.drdr.virginia.edu
       --> Education:The EcoGopher Environmental Library/
           --> Environmental Factsheets!/
                   --> EPA Chemical Substance Factsheets/
                       -->  4,4'-Thiodianiline.




Common Name:    4,4'-Thiodianiline
CAS Number:     139-65-1
DOT Number:     None
Date:           May, 1989
-----------------------------------------

HAZARD SUMMARY
* 4,4'-Thiodianiline can affect you when breathed in and by
  passing through your skin.
* 4,4'-Thiodianiline should be handled as a CARCINOGEN WITH
  EXTREME CAUTION.
* 4,4'-Thiodianiline can cause reproductive damage.
* 4,4'-Thiodianiline can affect the blood's ability to carry
  oxygen (methemoglobinemia) causing headache, dizziness,
  weakness and a blue color to the skin. Higher levels can
  cause shortness of breath, collapse and even death.
* Contact can irritate the eyes, skin, nose and throat.

IDENTIFICATION
4,4'-Thiodianiline is a needle shaped material. It is used as
.
.      <display deleted>
.
```

Reference Exercise 5. USE WAIS VIA GOPHER TO IDENTIFY A STANDARD

Task You are a structural engineer looking for specifications or standards regarding the construction of reinforced concrete drainpipes, and want to find any relevant reference source(s) on the Internet. You are most comfortable using Gopher.

Approach The very specific, detailed nature of this query calls for searching full-text documents using WAIS. Since you prefer to stick with Gopher, however, you will need to use a Gopher->WAIS gateway (that uses Gopher commands to search, retrieve and display WAIS sources).

Connect to a Gopher --> WAIS gateway available at Rice University.

```
% gopher chico.rice.edu
--> Other Gopher and Information Servers/
        --> WAIS (Wide-Area Information Servers)/
            --> Directory of WAIS servers <?>
```

First, search in the Directory of Sources (the master catalog of WAIS sources) for sources that index standards and specifications.

Words to search for

standards and specifications

Internet Gopher Information Client v2.0.16

Gopher returns a list of matching sources in order of relevance; the most relevant source should be ranked near the top of the list.

```
        1.  Information on database:  directory-of-servers
--->    2.  document_center_catalog.src <?>
        3.  OSHA-Standards.src <?>
        4.  PDS_standards.src <?>
        5.  internet-standards-merit.src <?>
        6.  internet-standards.src <?>
    .
    .   <text deleted>
    .
```

In the search box, clear out the terms from the previous search (e.g., 'standards and specifications') by pressing control-u; then enter new search terms.

Words to search for

reinforced concrete drain gutter

Internet Gopher Information Client v2.0.16

document_center_catalog.src: reinforced concrete drain gutter

Gopher returns a list of matching documents, in order of relevance.

```
--> 1.   ASTM-C1092 - GLASS REINFORCED CONCRETE D-LOAD..
    2.   ASTM-C655 - REINFORCED CONCRETE D-LOAD CULVERT..
    3.   ASTM-C655M - REINFORCED CONCRETE D-LOAD CULVERT..
.
.   <text deleted>
.
```

Here is a sample document. It is from the catalog of the Document Center, a fee-based technical document delivery service.

```
ASTM-C1092 - GLASS REINFORCED CONCRETE D-LOAD CULVERT, STORM
DRAIN, & SEWER94%
+------------------------------------------------------------+
ASTM-C1092 - GLASS REINFORCED CONCRETE D-LOAD CULVERT, STORM
DRAIN, & SEWER PIPE, STANDARD SPECIFICATIO

Stock status:              In stock
List Price:                23.00 + handling, shipping
(ground, overnight, 2nd
 day) and sales tax
Pages:                     5
Published by:              ASTM
Current edition effective: 07-15-94
Edition Status:            Current
Edition:                   1994 EDITION
Note:                      Prior editions exist, call for
information
Keywords:                  CONCRETE PIPE GLASS REINFORCED
CULVERT  D LOAD

To order call 415-591-7600, fax 415-591-7617, send email to
order@doccenter.com or send USmail to:

Document Center - 1504 Industrial Way, Unit 9 - Belmont, CA
94002

Orders received before 3:30pm are shipped the same day.


Mail current document to:

clementg@solix.fiu.edu
```

To save this entry for later reference, type '**s**' to save and then enter a name for the saved file (not shown here). To mail it to another user, type '**m**'and when prompted, type an email address.

Reference Exercise 6. USE WWW TO SEARCH THROUGH A REFERENCE TEXT

Task You are a geneticist who is interested in getting some background information on Tay-Sachs disease and would like to use the hypertext version of the well-known reference text on genetic disorders, *Online Mendelian Inheritance in Man.* How might you locate this text on the World-Wide Web?

Approach Since you already know the name of the desired reference source, it should be a simple matter to search for it using a WWW indexing tool, such as AliWeb.

Point your Web client (demonstrated here is the text-based Lynx) to the AliWeb page at Nexor.

NOTE: Type in the AliWeb URL directly after '**lynx**'. Here it is shown on two lines only because the string wouldn't fit on one line on this page!

Select the option for searching ALIWEB.

```
% lynx
http://web.nexor.co.uk/public/aliweb/aliweb.html

                    Welcome to ALIWEB (p1 of 2)
A Public Service provided by NEXOR

_____

                       WELCOME TO ALIWEB

    Welcome to ALIWEB, a resource discovery system for the WWW.
_____

*  Search  the  ALIWEB  database
*  Introduction to ALIWEB
*  Registering with ALIWEB
.
.       <display deleted>
.
.
_____


                    SEARCH THE ALIWEB DATABASE

    You can search the ALIWEB database in a number of ways:

        *  A  Search  Form  at  NEXOR

    At an ALIWEB Mirror site:
       * Indiana University (USA)
       * LEO Archive (Munich, Germany)
.
.       <display deleted>
.
.
_____

                       ALIWEB SEARCH FORM

    This form queries the ALIWEB database. You can provide
multiple search terms separated by spaces, and the results will
be displayed in a best-match order.
```

Enters search terms and click on the '**Submit**' key to run the search.

AliWeb returns a list of matches, in order of relevance. The top two entries are exact hits.

Select one of the links returned by AliWeb to jump to OMIM.

To run a keyword search, select the appropriate option.

In Lynx, the '**s**' key calls up a prompt for keywords. In a graphical Web client such as Mosaic, MacWeb, etc., clikcing on the '**?**' icon initiates a keyword search.

The keyword search retrieved multiple entries from the 'Online Mendelian Inheritance in Man'.

Click on the desired search result.

```
Search term(s): online mendelian inheritance Submit Reset
Search Results for 'online mendelian inheritance' (p1 of 6)

SEARCH RESULTS FOR 'ONLINE MENDELIAN INHERITANCE'

   [120] gdbwww.gdb.org
           The Genome Data Base (GDB) Web server at Johns
Hopkins University offers direct query access to the Genome
Data Base and Online Mendelian Inheritance in Man (OMIM).

   [120] OMIM
           Direct query access to OMIM, The Online Mendelian
Inheritance in Man -- the comprehensive source on genetic
disorders.

   [40] The Internet Plaza
           The Internet Plaza is dedicated to providing Internet
.
.     <display deleted>
.


                                        OMIM Top Level (p1 of 3)

                        OMIM SEARCH OPTIONS

       * OMIM using Fill-In Forms
        * OMIM using Keyword Search

       * Gene Map


This is a searchable index.  Use 's' to search

Enter a database search string: tay sachs


OMIM - Query Results (p1 of 10)

OMIM QUERY RESULTS

   OMIM IRX Query: tay sachs

       1. *272800 TAY-SACHS DISEASE [TSD; GM2-
GANGLIOSIDOSIS, TYPE I; B VARIANT GM2-GANGLIOSIDOSIS;
HEXOSAMINIDASE A DEFICIENCY; HEXA-; TAY-SACHS DISEASE,
JUVENILE, INCLUDED; HEXOSAMINIDASE A DEFICIENCY, ADULT TYPE,
INCLUDED; GM2-GANGLIOSIDOSIS, ADULT CHRONIC TYPE, INCLUDED;
TAY-SACHS DISEASE, VARIANT B1, INCLUDED; TAY-SACHS DISEASE,
PSEUDO-AB VARIANT, INCLUDED]
       2. *268800 SANDHOFF DISEASE [GM2-GANGLIOSIDOSIS, TYPE II;
0 @VARIANT GM2-GANGLIOSIDOSIS; HEXOSAMINIDASES A AND B
DEFICIENCY; HEXB-; HEXOSAMINIDASE B, INCLUDED; HEXB, INCLUDED;
SANDHOFF DISEASE, ADULT TYPE, INCLUDED; SANDHOFF DISEASE,
```

The first entry from the reference text is displayed. Note that each heading in the table of contents represents a hyperlink to a specific section, such as 'Diagnosis', 'Evolution', and so on.

In this case, select the link to Mini-MIM.

Once you find the bit of information you need, it can easily be saved to a file using the 'Print' option in Lynx, or the 'Save' option available in most graphical Web clients.

JUVENILE TYPE, INCLUDED]

OMIM -- 272800 (p1 of 107)

*272800 TAY-SACHS DISEASE [TSD; GM2-GANGLIOSIDOSIS, TYPE I; B
VARIANT GM2-GANGLIOSIDOSIS; HEXOSAMINIDASE A DEFICIENCY; HEXA-;
.
.   <display deleted>
.

TABLE OF CONTENTS

        * **Mini-MIM**
        * Description
        * Phenotype
             + Clinical Features
             + Biochemical Features
        * Genotype
             + Mapping Information
             + Molecular Genetics
        * Diagnosis
        * Population Genetics
        * Evolution
.
.   <display deleted>
.

MINI-MIM

Tay-Sachs disease, an autosomal recessive, progressive
neurodegenerative disorder, which in the classic infantile
form, is usually fatal by age 2 or 3 years, results from
deficiency of the enzyme hexosaminidase A (Hex-A). Juvenile and
adult-onset forms exist.
.
.   <display deleted>
.
 H)elp O)ptions P)rint G)o M)ain screen Q)uit /=search
[delete]=history list **P**

            Lynx Printing Options

            PRINTING OPTIONS

   There are 2121 lines, or approximately 33 pages, to print.
   You have the following print choices
   please select one:

   Save to a local file

Module 5: Searching (and Retrieving) the Literature

IN THIS MODULE:

Introduction
What's available
Quality and value
Access methods
Finding bibliographic and document delivery resources
Exercises

1. *Use HYTELNET to find and connect to a bibliography*
2. *Find a bibliography starting from zero*
3. *Find and download a WAIS source for patent searching*
4. *Find a journal table of contents*
5. *Find and search a journal publisher's Web site*
6. *Conduct a full-text search of literature reviews*

Introduction

Effective access to the published literature—so essential to providing researchers and practitioners the foundation for moving forward—requires two categories of resources: (1) bibliographic tools to identify relevant materials, and (2) document delivery systems to retrieve them. Bibliographic tools facilitate both retrospective searching and current awareness, and include periodical indexes and abstracts, current awareness services, subject-specific bibliographies, and library catalogs. Resources for retrieving the literature include personal subscriptions, protocols for reprint requests, library journal holdings, interlibrary loan services, and fee-based article ordering services. Well-developed, effective resources in both categories have been in wide use within the scientific community long before the Internet emerged as a significant research tool.

The Internet has brought the traditional search and retrieval tools to researchers' desktops. In response to the increasing acceptance of (and preference for!) Internet-accessible research tools, scientific publishers, database producers and information providers are quickly setting up Internet nodes and developing Internet-based services to add value to existing service and products and develop new markets. Libraries are also now undertaking large-scale digitizing projects of sci/tech materials, in support of national and regional initiatives funded by the National Science Foundation, NASA, the Department of Defense, and other science-related organizations.[1]

As the amount of primary sci/tech information available via global networks continues to expand, it is becoming increasingly important for scientists, practitioners, and other users of sci/tech information to understand how to use and apply Internet tools to search and retrieve the literature.

 16

What's Available

Abstract and indexing services

A large number of established abstract and index producers are making their products accessible over the Internet as a value-added service. Internet users may connect to these services through the information provider's own Internet node; through an existing online system, such as the commercial online databanks STN (Chemical Abstract Services) or RLIN (Research Libraries' Group); or through a gateway on a library's online public access catalog (OPAC). Keep in mind that Internet access to commercial bibliographic databases does not eliminate the search charges or contractual agreements that may apply to these services (information providers still need to recover costs, with or without the Internet). Usernames, secure passwords, and monthly billing may all be required.

Online databanks offering full-text (and graphics) output formats

Commercial online services that provide access to bibliographic databases are moving to provide options for full-text (and graphics) output display, essentially merging the process of bibliographic searching and document retrieval. Users of Chemical Abstracts' STN online service, for example, may search the full text and display the complete article from several dozen American Chemical Society journals. Access to STN over the Internet makes it possible for researchers to find and retrieve information on demand from one's own office or laboratory.

Specialized bibliographies

For nonprofit or smaller producers of bibliographic databases, *e.g.*, libraries, government agencies, individuals or small research organizations, the Internet provides an efficient, cost-effective means for providing global access to their bibliographic databases. What may have started as a stand-alone, local resource—a library catalog, a specialized bibliography, or a technical report collection—may gain use among researchers worldwide once loaded on an Internet-accessible node.

Specialized bibliographic databases often provide extensive, in-depth access to the literature of a specialized field. For example, the Sea Turtles bibliography produced by the Archie Carr Center for Sea Turtle Research at the University of Florida is a unique, comprehensive source for citations to articles and other background information on that subject. The Buckyball Database produced by the Department of Physics at the University of Arizona covers the literature of fullerenes and related chemical structures. Without Internet access to these databases, most researchers would have difficulty taking advantage of these resources.

Current Awareness

Another class of bibliographic resources finding popular application on the Internet are the current awareness services providing tables of contents and sometimes abstracts for current or future issues of scientific journals. Those providing current awareness services include scientific publishers such as Springer-Verlag and Elsevier, and scientific societies such as the American Astronomical Society. The popular UnCover table of contents service initiated by the Colorado Alliance of Research Libraries (CARL),[2] developed out of the efforts of libraries to enhance public access to their journal holdings.

Often coupled with the current awareness service is an article ordering feature, offering printed copies of articles faxed or mailed for a fee. Upon ordering selected articles found during a search, users provide payment from either a deposit account or credit card.

Document Delivery Services

As publishers seek (and users demand) more cost-effective and efficient means of retrieving scientific publications online, a number of prototype document delivery services are appearing on the Internet, providing electronic copies of printed articles on demand. Sponsored by partnerships between commercial publishers, scientific societies, universities or libraries, these experiments in electronic distribution of publications are testing a variety of issues related to large-scale electronic distribution of scientific articles: presentation and distribution methods, user acceptance, mechanisms for copyright protection and cost recovery, and so on.

Well known among such experiments is Elsevier's TULIP project, involving Internet distribution of several dozen journals in materials science to selected universities.[3] A similar project in the nonprofit sector is STELAR (Study of Electronic Literature for Astronomical Research), sponsored by NASA, the American Astronomical Society, with support from research libraries, publishers, and other organizations.[4]

Internet document delivery systems remain distinct from electronic publication, providing access to the printed literature rather than material designed specifically for Internet presentation, storage and delivery. For further discussion about this latter subject, see the next module, "Electronic Publishing on the Internet."

 17

Quality and Value

For bibliographic access tools in any format, whether Internet, CD-ROM, or printed, quality and value relate to the following aspects:

>Scope (comprehensive or narrow coverage of the subject)

>Selection process (source materials covered)

>Format (structure of database; number of type of fields available; search and display features)

>Currency (frequency of updates; lag time of coverage)

Established services produced by well-known and reputable producers meet these standards. What's more, their value is enhanced by enabling researchers to perform searches any time, any where, from their own network-connected computers.

For the services provided by smaller producers, however, users may need to carefully evaluate the quality of the service. Is the bibliographic data accurate? Is coverage spotty or complete? Is the database routinely updated?

As with any Internet resource, the disadvantages of the smaller, less established bibliographic databases need to be weighed against its potential usefulness. Does it focus in depth on the subject of interest to the searcher? Does it include useful materials (technical reports and other grey literature) excluded from the more traditional services?

For document delivery resources, quality and value relate largely to the visual clarity of the material provided. Photographs, maps, and other graphics must be as detailed and clear as those provided in the printed version of the journal. Speed of delivery and cost of the requested documents are also considerations that may determine whether the user (or library) retrieves the documents over the Internet, or uses alternative access methods.

 18

Access Methods

Most bibliographic tools on the Internet, including indexes, abstracts, current awareness services, and specialized bibliographies, are interactive services accessible with the Telnet protocol. Users may either connect to such services with the Telnet command either directly from the command line, or through a menu selection on a library catalog, Gopher menu or Web Page.

Some bibliographies are also searchable using WAIS, a tool capable of performing a simple full-text search on the journal articles. However, the limited search options of many WAIS clients (*e.g.,* no proximity connectors, field-specific searching, or limiting by date) severely limit the effectiveness of WAIS as a retrieval tool.

Smaller bibliographies may also be available as text files retrieved in their entirety using electronic mail, FTP, Gopher, WAIS or the World-Wide Web. This means of access provides limited ability to search the document for needed information.

Document delivery services on the Internet may require a wide range of access methods, as publishers experiment with a variety of formats for capturing printed pages in electronic format and then delivering and storing them online.

The two most common formats for electronic distribution of science journals are full-text ASCII files or scanned (bitmapped) images of journal pages. ASCII format provides no graphics, but is searchable; scanned page images may convey high-quality graphics, but are not searchable. In many instances, publishers are using a combination of both formats to produce an effective search-and-retrieval service— searchable text abstracts, combined with high-quality pages for display or output.

Finding Bibliographic and Document Delivery Resources

Commercial database services on the Internet are easy to find because their producers tend to advertise them widely. They are likely to provide 800- help lines to answer questions and help prepare search strategies. Also, libraries and other research institutions are likely to subscribe to the commercial online services, and may have a trained staff member available to perform a mediated search, or help the user prepare one.

Noncommercial databases, such as the specialized bibliographies or the current awareness service UnCover, may be less well known and often hard to find. They may be hidden as a menu item on a particular library catalog or gopher server. For example, the Sea Turtles Bibliography appears on a menu of LUIS, the online catalog for the State University System of Florida. The Buckyball Database is reached through the online catalog for the University of Arizona. Since most of these data-bases are likely to be accessible by Telnet, one may locate them with HYTELNET, the Internet search tool for Telnet-accessible resources.

Another approach to finding search and retrieval services on the Internet is to look for the Internet site of the producer. Expect that the leading scientific publishers and database providers have established a presence on the Internet, most likely as a gopher or Web site. Using what you know about the producer's or publisher's name and geographic location, you may track them down through geographically-arranged such as the "Around the World" hierarchy on Gopher

```
gopher://gopher.tc.umn.edu/11/
Other20%Gopher20%and20%Information20%Servers/
all
```

or the "W3 Servers List" arranged by geographic location

```
http://info.cern.ch/hypertext/DataSources/WWW/
Servers.html
```

For more guidance in finding resources in your subject area, see Section C of this guide.

Endnotes

1. Some of the most ambitious of these projects fall under the umbrella of the Digital Library initiative, in which select U.S. research libraries have been funded to develop prototype digital collections and information retrieval systems for environmental science, earth and space science, engineering, geographic information, and more. Each of these projects is summarized in the article: Cole, Tim and Susan Harum, "Digital Library Projects," *LITA Newsletter*, vol. 16, no. 2 (Spring 1995): 25-27.

2. The UnCover service is available at

    ```
    telnet://database.carl.org
    ```

    Users may search tables of contents at no charge, and may establish a user profile or deposit account for online ordering of articles.

3. For more information on the TULIP project, see Elsevier's Internet service at

    ```
    http://www.elsevier.nl
    ```

    or send mail to

    ```
    nlinfo-f@elsevier.nl.
    ```

4. For more information about STELAR, send mail to

    ```
    stelar-info@hypatia.gsfc.nasa.gov.
    ```

Literature Exercise 1. USE HYTELNET TO LOCATE AND CONNECT TO A TELNET-ACCESSIBLE BIBLIOGRAPHY

Task As an engineer, you are interested in searching for articles and reports, and heard about Quakeline, a searchable bibliographic database for earthquake engineering. Where is this service and how can you connect to it?

Approach Bibliographic databases such as library catalogs and bibliographies tend to be accessible by Telnet, and the best tool for finding Telnet sites is HYTELNET.

Connect to one of the publicly-accessible HYTELNET services at Washington & Lee University.

```
% gopher  liberty.uc.wlu.edu
--> Explore Internet Resources/
      --> Telnet Login to Sites (Hytelnet)/

               Telnet Login to Sites (Hytelnet)
```

Since you already know the name of the database, you may conduct a keyword search on its name.

```
-->   1.  Hytelnet (word search) <?>
      2.  Library Catalogs /
      3.  Library Catalogs, By System Type /
      4.  Library Catalogs, Help Files /
      5.  New or Revised Hytelnet Entries/
      6.  Other Resources /
+--------------------------Hytelnet (word search)------------+
| Words to search for
|
| quakeline
|
| [Help: ^-]  [Cancel: ^G]
+-----------------------------------------------------------+


-->   6.  Other Resources /

                   Other Resources

      1.  Archie: Archive Server Listing Service/
 -->  2.  Databases and bibliographies/
      3.  Distributed File Servers (Gopher/WAIS/WWW)/
      4.  Electronic books/
      .
      .

               Databases and bibliographies

      .
      .
      55. Physicians' GenRx International (tm)/
      56. Project Hermes (Supreme Court Decisions)
```

If you did not know an exact name for the database, you could also find it by selecting the appropriate menu and browsing until you find what you want.

Here's the bibliography.

HYTELNET returns the matching menu item.

```
 -->  57. QUAKELINE, National Center for Earthquake Engine.../
      58. RAPID: ESRC database of Research Abstracts.../
      .
            <display deleted>
```

Note that it is a Telnet
connection, as denoted by
the <TEL> at the end of
the Gopher menu line.

A warning screen tells you
what Telnet host you are
going to and how to close
the connection if you get
stuck.

Telnet usually requires the
user to input the desired
emulation mode. VT100 is
most common.

As with many
bibliographic databases,
Quakeline resides on a
computer system that
houses the site's library
catalog, along with other
databases and indexes.

Note that the 'Welcome'
screen provides instructions
on how to exit the system
(STOP). You should make a
written note of this
command -- it may not
appear later when you need
to use it.

Note the asterisk (*),
denoting that some
databases on this library
system are password-
controlled and not
available to the general
public.
As with any Telnet-
accessible site on the

```
QUAKELINE, National Center for Earthquake Engineering Research

--->   QUAKELINE, National Center for Earthquake Engin... <TEL>



+-QUAKELINE, National Center for Earthquake Engin.. Research-+
|
| Warning!!!!!, you are about to leave the Internet
| Gopher program and connect to another host. If
| you get stuck press the control key and the
| ^ key, and then type q.
|
| Connecting to bison.cc.buffalo.edu, port 23 using telnet.
|
|        [Cancel: ^G] [OK: Enter]    |
+------------------------------------------------------------+


ENTER TERMINAL TYPE: vt100

                        Welcome to BISON
                 Buffalo Information System Online

                     DATABASE SELECTION MENU

 To select a database, type a LABEL and press RETURN.  Enter
START  to  return to the main menu.  Enter STOP to exit the
system.
        1               UB Libraries Catalog
        2               Other Catalogs
        3               Business / Legal Indexes
        4               Education / Library Science Indexes
        5               General Indexes
        6               Humanities / Art / Social Science Indexes
        7               Science / Technology Indexes


Database Selection: 7


                     DATABASE SELECTION MENU

 To select a database, type a LABEL and press RETURN.  Enter
START  to return to the main menu.  Enter STOP to exit the
system.
                 SCIENCE / TECHNOLOGY INDEXES
   *SCII    Wilson Science and Technology Indexes 1989+
                 (SCII includes:  Applied Sci & Tech Index;
                 Bio and Agri Index; and General Science Index.)
   QKLN          QUAKELINE (Earthquake Resources) 1987+

* Databases that require Sign-On.
  Database Selection: qkln
```

Internet, the commands and options for searching this database are not 'Internet' commands, but rather, specific commands for this system.

You need to read the screens carefully to find out how to search for and display information.

A subject search retrieves 205 entries.

Introduction

QUAKELINE , the database of the National Center for Earthquake Engineering Research (NCEER), contains bibliographic information on the subjects of earthquakes, earthquake engineering, natural hazards mitigation, and related topics. The time period covered by the QUAKELINE database is 1987 to the present with selected coverage of documents published prior to 1987.

QUAKELINE contains citations for journal articles, monographs, conference papers, technical reports, maps, data compilations, pamplets, videotapes, and slides. Most citations include abstracts, subject terms ("identifiers") and information that may be used to acquire documents.

```
TO SEARCH BY: Title    ENTER:   t=reinforced concrete structure
              Author            a=jirsa j
              Subject           s=concrete dams
              Keyword           k=hazard? and mitigation

   NEXT COMMAND: s=reinforced  concrete  columns
```

Search Request: S=REINFORCED CONCRETE COLUMNS   QUAKELINE
 Search Results: 205 Entries Found
Subject Guide

```
LINE:  BEGINNING ENTRY:                        INDEX RANGE:
 1     REINFORCED CONCRETE COLUMNS             1 -   15
 2     REINFORCED CONCRETE COLUMNS             16 -   30
 3     REINFORCED CONCRETE COLUMNS             31 -   45
            .
            .
            .
13     REINFORCED CONCRETE COLUMNS             181 - 195
14     REINFORCED CONCRETE COLUMNS             196 - 205
```

STArt over   Type number to begin display within index range
HELp
OTHer options

NEXT COMMAND: 1

Search Request: S=REINFORCED CONCRETE COLUMNS   QUAKELINE
Search Results: 205 Entries Found
Subject Index

```
      REINFORCED CONCRETE COLUMNS
  15    BEHAVIOR OF REINFORCED CONCRETE COLUMNS SUBJ <1986>
  16    BEHAVIOR OF REINFORCED CONCRETE COLUMNS UNDE <1978>
  17    BEHAVIOR OF REINFORCED CONCRETE COLUMNS UNDE <1986>
  18    BEHAVIOR OF STRENGTHENED AND REPAIRED REINFO <1985>
```

```
            19    BEHAVIOR OF THE CONNECTION OF R C COLUMN WIT <1992>
            20    BEHAVIOUR OF R C COLUMNS UNDER STATIC COMPRE <1991>
            21    BEHAVIOUR OF REINFORCED CONCRETE POLES DURIN <1990>
            22    BEHAVIOUR OF REPAIRED STRENGTHENED REINFORCE <1990>
            23    BEHAVIOURS OF REINFORCED CONCRETE COLUMNS UN <1987>
            24    BIAXIAL COLUMN ELEMENT OF NONLINEAR DYNAMIC <1991>
            25    COLUMN SAFE DESIGN METHOD OF MULTI STORY BUI <1989>
            26    COMPARATIVE STUDY OF PROBABILISTIC METHODS I <1994>
                  .
                  .
    NEXT COMMAND: 26
```

Choose a record to display.

```
    QUAKELINE Record -- 26 of 205 Entries Found
    Brief View
    -------------------------------------------------------------
    AUTHOR (AU)     Sellier-A.
                    Lorrain-M.
                    Pinglot-M.
                    Mebarki-A.

     TITLE (TI)     COMPARATIVE STUDY OF PROBABILISTIC METHODS IN
    STRUCTURAL SAFETY ANALYSIS.

     SOURCE (SO)    Structural Safety and Reliability: Proceedings
    of ICOSSAR '93, The 6th International Conference on Structural
    Safety and Reliability; Innsbruck, August 9-13, 1993.  A A
    Balkema, Rotterdam, 1994, volume 2, pages 1329-1336.

     NOTES  (NT)     23 references.  Figures, graphs, tables.
    Sixth International Conference on Structural Safety and
    Reliability.
    ---------------------------+ Page 1 of 2 -------------
     STArt over     HOLdings       GUIde      <F8>  FORward page
     HELp           LONg view                 <F6>  NEXt record
     OTHer options  INDex                      <F5>  PREvious record


    NEXT COMMAND: for
```

Type 'for' to see the second screen of this display.

```
    Search Request: S=REINFORCED CONCRETE COLUMNS
    QUAKELINE (Earthquake R
     QUAKELINE Record -- 26 of 205 Entries Found
    Brief View
    -------------------------------------------------------------
    COMPARATIVE STUDY OF PROBABILISTIC METHODS IN STRUCTURAL
    SAFETY ANALYSIS...
      (CONTINUED)

    LOCATION (LO)    SEL TA656.5.I57 1993 v.2.

    ---------------------------+ Page 2 of 2 -------------

    NEXT COMMAND: stop
```

When you are ready to exit the system, type 'stop' to logoff. This returns you to the Gopher menu for Hytelnet.

Literature Exercise 2. FIND A BIBLIOGRAPHY STARTING FROM ZERO

Task You are a chemist looking for a recent paper by Heiney on the chemical structure of C70, but can't remember where it's published, or anything else about it for that matter! Maybe you can identify it by browsing the Buckyball Database, a bibliography of background material and current citations on the topic of fullerenes and related structures. But you don't know where to find this database on the Internet.

Approach You need to track down the address for the Buckyball database without knowing what kind of server it resides on or the exact name of the database. You might have success using Veronica, which allows keyword searching on partial search terms. Even if BuckyBall Database is a Telnet-accessible catalog, it may be available through a Gopher gateway.

Veronica is found on most Gopher menus, but is also accessible directly from the 'mother site' at the University of Nevada.	`% gopher gopher.unr.edu`
	
	`-->Search ALL of Gopherspace (4800 servers) using Veronica/`
	
	`          Internet Gopher Information Client 2.0 pl10`
	
	`   Search ALL of Gopherspace (4800 servers) using Veronica`

From the list of Veronica servers, select a server. Since you are 'shooting in the dark', it is also a good idea to select the search type "Search GopherSpace by Title Words" to get the maximum possible results.	`      1.How to Compose Veronica Queries - June 23, 1994.`
	`      2.Frequently-Asked Questions (FAQ) about Veronica....`
	`      3.About Veronica:  Documents, Software, .../`
	`      .`
	`      .    Display deleted>`
	`      .`
	`-->   10.Search GopherSpace by Title word(s)(via U.of Pisa) <?>`
	`      11.Search GopherSpace by Title word(s) (via SUNET) <?>`
	`                 Page: 1/2              Internet Gopher`

In the search box, enter known portion of search string, truncating with an asterisk (*).	`Words to search for: buckyball*`

Veronica returns the search results in the form of a Gopher menu.	`Search GopherSpace by Title word(s) (via University of Pisa):`
	`buckyball*`
	
	`      1.Buckyball Nomenclature.`
	`      2.CCL:Administrator of the Buckyball database?.`
	`      3.Buckyball Database (ftp)/`
	`      4.Instructions on using the Buckyball Database.`
	`      5.#0123: BUCKYBALL MOLECULAR FUSION .`
	`      6.CCL:Administrator of the Buckyball database?.`
	`      7.buckyball.ray.`

Eureka! Here's a Telnet link to the database.	`-->  8.Buckyball Database (Univ. of Arizona Library) <TEL>` `      9.1st Practical Applications Of "Buckyballs" Seen...`

To bookmark this resource so you can return to it again, press 'a' while the arrow is pointing to it.

To obtain its direct address, press '=' (the equal sign).

'Type 8' denotes a Telnet resource.

'Path' info indicates direct telnet instructions.

From the Veronica menu, select the Telnet connection to Buckyballs Database.

A message then appears indicating that you are about to leave Gopher and connect to sabio.arizona.edu, port 23 using Telnet.

The welcome screen for SABIO at the University of Arizona then appears.

An introduction to the Buckyball Database appears, with a list of command options.

.

Enter the option for a keyword search, and then enter the search term.

```
Link Info (0k)
+------------------------------------------------------------+
#
Type=8
Name=Buckyball Database (University of Arizona Library)
Path=sabio; select Other databases then Buckyballs
Host=sabio.arizona.edu
.
.    <Display deleted>
.


 +-------Buckyball Database (University of Arizona Library)---+
 | Warning!!!!!, you are about to leave the Internet          |
 | Gopher program and connect to another host. If             |
 | you get stuck press the control key and the ]              |
 | key, and then type quit                                    |
 |                                                            |
 |      Connecting to sabio.arizona.edu using telnet.         |
 |                                                            |
 | Use the account name "sabio; select Other databases then   |
 | Bucky" to log in                                           |
 .
 .      <Display deleted>
 .
```

Select "**OTHER databases and remote Libraries**";
then "**Buckyball Database**"


```
Sabio presents the Buckyball Database in collaboration with the
University of Arizona Physics Dept. It covers fullerenes,
containing citations to published as well as submitted
articles.

   You may search for articles by any of the following:

                K > KEYWORD (in title)
                A > AUTHOR
                T > Article TITLE
                J > JOURNAL Title
                I > Buckyball INFORMATION
                G > Change Language/CAMBIO A ESPANOL
                R > RETURN to the library catalog

                Choose one (K,A,T,J,I,G,R) k

                    WORD : c70
```

256 entries found, entries 1-8 are:

1 13C NMR Study Of Orientational Ordering In S Solid
2 2D Nuclear Magnetic Resonance Study Of The S
3 A Collisional Study Of Some C60 And C70 Full Rapid 4 A
Model For The C60 And C70 Growth Mechanism

.

.   <display deleted>

.

Please type the NUMBER of the item you want to see, OR
F > Go FORWARD   D >DISPLAY Title and Author
N > NEW Search   L >LIMIT this Search

Choose one (1-8,F,N,A,D,L,J,E,O) **L**

You may limit your search by any of the following

                Y > YEAR of Publication
                T > Words in the article TITLE
                A > Words in the article AUTHOR
                J > Words in JOURNAL title
                R > RETURN to Browsing

                Choose one (Y,T,A,J,R) **A**

        Words in the article AUTHOR: **heiney**

 4 entries found, entries 1-4 are:

1 Molecular Orientation Dynamics In Solid C70:
2 Orientational Disorder In Solvent-Free Solid
3 Phase Transitions In Solid C70: Supercooling

.

.   <display deleted>

.

Choose one (1-4,R,N,A,D,E) **3**

You searched for the WORD: c70LIMITED TO Words in the article
AUTHOR has "hei

 AUTHOR   McGhie, A.R.
                Fischer, D.
                Heiney, P.A.
                Stephens, P.W.
                Cappelletti, R.L.
                Neumann, D.A.
                Mueller.
                Mohn.
                Ter Meer.
 TITLE   Phase Transitions In Solid C70: Supercooling,
Metastable Phases And Impurity Effect.
 JOURNAL   Physical Review B - Condensed Matter.
 SUBMITTED   1/31/94.

Too many entries are found.

Choose the option to the further limit the search by author.

Number 3 is the desired article.

You may then print out the citation using the **<print screen>** key, To continue the search, or exit the system choose "**N**" for New Search;or "**R**" to Return to Library Catalog; then "**E**" to exit.

This will close the Telnet connection and return you to the Gopher session.

Literature Exercise 3. USE GOPHER TO FIND AND DOWNLOAD A WAIS SOURCE FOR PATENT SEARCHING.

Task As a materials scientist, you are interested in finding recent patents for a process to produce zirconium silicate. Can you conduct a patent search over the Internet?

Approach The U.S. Patent and Trademark Office has an Internet site at the Internet Town Hall. Maybe that site provides information about searching patents on the Internet.

Connect to the Town Hall Gopher.

Select the appropriate directory.

```
% gopher town.hall.org

        Internet Gopher Information Client 2.0 p110

           Root gopher server: town.hall.org

        1.   Welcome to the Internet Town Hall.
        2.   Access methods available for the Town.../
        3.   Federal Reserve Board/
        4.   General Services Administration/
        5.   SEC EDGAR/
  -->   6.   U.S. Patent and Trademark Office/
```

Always a good idea to read the 'About' file first.

```
        Internet Gopher Information Client 2.0 p110

           U.S. Patent and Trademark Office

  -->   1.   About the Patent Full-Text/APS Distribution.
        2.   Keyword Search of the U.S. Patent...Data <?>
        3.   WAIS source description for Patent index.
        4.   Help on performing WAIS searches.
```

In consideration of this recommendation, it looks like WAIS will be a better means of access.

The WAIS source file for the patents is a special file, patent.src, available on this Gopher. But it is not useable on this site -- it must be downloaded and used with your own WAIS client (not a public client).

```
About the Patent Full-Text/APS Distribution (0k)
100%
+-------------------------------------------------------+
This subdirectory contains Full Text Patent Data for
1994.The data1 subdirectory is organized by ranges of
patent numbers.  We *highly* recommend that you use
WAIS to access this information.

Transfer the file patent.src back to your home system
and put it with your other WAIS source files.
+-------------------------------------------------------+
[Help: ?]                [Exit: u] u
```

Read the source description
for further info.

The information at the top
is used by your WAIS client
to determine what server to
connect with and which
database to search.

To save the file, use
Gopher's '**s**' option.

Or, you may download the
WAIS source by moving the
arrow to the appropriate
line on the Gopher menu
and pressing '**D**' to
download.

In the query box, indicate
which transfer protocol to
use for the download.

```
-->  3.  WAIS source description for Patent index


WAIS source description for Patent index (0k)
100%
+------------------------------------------------------+
(:source
   :version  3
   :ip-address "192.101.98.5"
   :ip-name "town.hall.org"
   :tcp-port 210
   :database-name "patent"
   :cost 0.00
   :cost-unit :free
   :update-time (:time-interval :interval :weekly :day
      5 :hour 1 :min 30 )
   :maintainer "waismaster@town.hall.org"
   :description
"Patent Full-Text/APS File for 1994.  Field name
abbreviations in the original feed have been expanded
.
.    <display deleted>
.


   1.   About the Patent Full-Text/APS Distribution
   2.   Keyword Search of the U.S. Patent...Data <?>
 --> 3.   WAIS s
+---WAIS source description for Patent index--+
|                                             |
|  -->      1. Zmodem                         |
|           2. Ymodem                         |
|           3. Xmodem-1K                      |
|           4. Xmodem-CRC                     |
|           5. Kermit                         |
|           6. Text                           |
|                                             |
+---------------------------------------------+
```

*   *   *   *   *   *

NOTE: Once the source file 'patent.src' is downloaded to the computer running your
WAIS client, you need to open WAIS, select the option for 'New' or 'Edit' source
(depending on the specific client you have) and select the file 'patent.src'. After
viewing and saving the document in WAIS, 'patent.src' should appear on the list of
sources available from your WAIS client, and may then be selected for use with a
query.

Literature Exercise 4. FIND A JOURNAL TABLE OF CONTENTS WITH USENET

Task You are a biologist interested in finding out what journals are making their table of contents available over the Internet.

Approach Many publishers are posting their tables of contents to one or more electronic discussion groups. For biologists, in particular, a Usenet newsgroup has been set up to serve as a clearinghouse for this purpose (in other subject fields, look for discussion groups oriented to the particular subject(s) covered by the journal).

Start the newsreader tin.

The newsreader creates the user's .newsrc file (or reads an existing one if this is not the user's first Usenet news session).

The 'Group Selection' screen indicates that the newshost makes almost 7000 newsgroups available to Usenet users at this institution. It also displays an alphabetical list of newsgroups.

Biology-related groups will fall under either the 'bionet.*' or 'sci.*' hierarchies.

You may search for the first occurrence of 'bio' by typing '/' to run a search on newsgroup names.

The menu jumps to the first of the biology-related newsgroups. The 'u' indicates that you are currently unsubscribed; the next column indicates the number of the group.

With the desired group selected, press **\<return\>** to see what tables of contents are available. The message 'no articles'

```
% tin
tin 1.2 PL2 [UNIX] (c) Copyright 1991-93 Iain Lea.
Connecting to newshost...
Reading news active file...
Reading attributes file...
Reading newsgroups file...

          Group Selection (newshost  6791)
h=help

        17     1   alt.angst.xibo.sex
        18    51   alt.appalachian
        19  1778   alt.aquaria
        20    86   alt.archery
        21    48   alt.architecture
        22    28   alt.artcom
        23    87   alt.asian-movies
.
.      <display deleted>
.


Search forwards> bio

.
.      <display deleted>
.


u 2349      -      bionet.jobs
u 2350      -      bionet.journals.contents
u 2351      -      bionet.molbio.ageing
u 2352      -      bionet.molbio.bio-matrix

bionet.journals.contents (0T 0A 0K 0H R)
h=help


            *** No Articles ***
```

indicates that no table of contents are currently available from bionet.journals.contents.

Returning to tin later, you may specify the desired newsgroup at start up.

The Group Selection screen indicates that there are now two tables of contents available for the newsgroup.

The '**u**' indicates that you are not subscribed to the group. To subscribe type an '**s**'.

The menu lists the journal titles for which a current table of contents are available, and the person who posted the message.

The table of contents provides basic information about the journal, subscription instructions, indexing details, and article citations.

To save the news article as a file, type '**s**', then '**a**'.

Name the file.

Choose a processing option, if desired. Since this is a short text file, no compression or archiving is needed.

Repeat steps above to save the next table of contents file. Then type '**q**' to quit.

```
%  tin  bionet.journals.contents
tin 1.2 PL2 [UNIX] (c) Copyright 1991-93 Iain Lea.
Connecting to newshost...
Reading news active file...
Reading attributes file...
Reading newsgroups file...


Group Selection (newshost  1)

   u   1     2  bionet.journals.contents



                   *** End of Groups ***

Subscribed to bionet.journals.contents



1 + Binary, vol. 6, no. 6, December 1994   BIOSCI
Administrat
2 + Journal of Cancer Research and Clinical Oncol
BIOSCI Administrat



CC BINARY (1994) Volume 6 Issue 6
CC DECEMBER 1994
.
.      <display deleted>
.
AU Lloyd-D.
TI Book Reviews: The Rainbow and the Worm:
   the Physics of Organisms
SO Binary-Comput-Micro.  1994 Dec.  6(6).  P 188.
.
.      <display deleted>
.
Save a)rticle, t)hread, h)ot, p)attern, T)agged
articles, q)uit: a

Save filename []> binary.toc

Process n)one, s)har, u)ud, l)ist zoo, e)xt zoo, L)ist
zip, E)xt zip, q)uit: q

         -- 1 Article(s) saved --
```

Literature Exercise 5. FIND AND SEARCH A JOURNAL PUBLISHERS' WWW SITE

Task   You are an electrical engineer needing to check recent issues of *Signal Processing* to see if an expected article is in print yet. The journal is not held at your library, and you need it immediately.

Approach *Signal Processing* is a title from Elsevier, a large, well-known scientific publisher in the Netherlands. It is likely they maintain an Internet site that will indicate if/where any of their current awareness services are available.

The WorldWide Web is your preferred means of access, so you can start looking for Elsevier in the geographically-arranged listing of Web sites, maintained at CERN, the WWW mothersite.

The opening page of the 'World List of W3 Servers' begins an extensive list of servers, starting with Africa and moving to each continent and country in alphabetical order.

Moving from one screen to the next with the `<spacebar>` will take a long time to get down to the Netherlands section.

As a shortcut, press the '/' key in Lynx (or click on the ? icon in a graphical Web browser such as Mosaic) to input a keyword.

```
% lynx   http://info.cern.ch/hypertext/DataSources/WWW/
(Do not actually insert linebreak!)          Geographical.html


                    WORLD LIST W3 SERVERS

   This is a list of registered WWW servers alphabetically by
continent country and state. ( About this list )

   New: A summary of the list is available

   See also: data available by other protocols , data by
subject , how to make a new server , test servers ,
automatically collected list of Home Pages, What's New, and the
clickable world map . If servers are marked "experimental", you
should not expect anything. Please see how to send
announcements of new servers (or modify your server's
description).

Africa

  SOUTH AFRICA

     Pretoria

   UNISA (University of South Africa)
          Information about study at Unisa. Also information
regarding on-campus Internet services, a WWW Library, other WWW
sites in South Africa, the WWW Worm, Archieplex, mirrors and
many more.

Enter a search string: netherlands
```

From that point, browse for the desired entry.

```
NETHERLANDS

    Ajax

  History and other information about the Ajax football
        team

  AT Computing
        Center for UNIX education and training

  CAOS, Center for computer aided chemistry and
        bioinformatics
  .
  .     <text deleted>
  .
    Elsevier Science
```

Here it is! Highlight it and press <**return**> to select.

Elsevier's Web site provides information for subscribers, resources for public users, and more.

Here is a link to a current awareness service for engineers.

```
              Elsevier Science - Home Page (p1 of 3)

                    ELSEVIER SCIENCE

            General Information from Elsevier Science

* The Elsevier Science Internet Journal and Book Catalogue
* Information on TULIP - The University Licensing Program
* About Elsevier Science

Electronic Journals

* Computer Networks and ISDN Systems: WWW '94 Conference
Proceedings

Current Awareness and Archival Services

* ECONbase
* Electrical and Electronic Engineering - Alert
* Mathematics & Computer Science - Alert
* Nuclear Physics Electronic

-- press space for next page --
DaArrow keys: Up and Down to move. Right to follow a link; Left
to go back.

 H)elp O)ptions P)rint G)o M)ain screen Q)uit /=search
[delete]=history list
```

This link is actually a Gopher server. You could have also found this site browsing the 'Gophers Around the World' hierarchy, or by running a keyword search in Veronica.

Select the desired journal title.

```
Select one of:

ELECTRICAL AND ELECTRONIC ENGINEERING ALERT (PUBLIC SECTION)
        (FILE) About the EEE-Alert Gopher
        (FILE) About the EEE-Alert contents service
        (FILE) About the EEE-Alert abstracts service
        (FILE) Copyright Notice & Disclaimer
J O U R N A L S
        (DIR)  Signal   Processing
        (DIR) Image Communication
        (DIR) Speech Communication
M I S C E L L A N E O U S
        (DIR) Subscriber Section (abstracts, papers, search)
        (DIR) How to get access to the Subscriber Section
        (DIR) Top level of Elsevier Science Gopher
        (DIR) LaTeX Styles
        .
        .       <text deleted>
        .
```

You may now browse some table of contents for the journal.

```
Signal Processing

                        SIGNAL PROCESSING

P U B L I C - S E C T I O N
        (DIR) Signal Processing, catalogue information
        (FILE) Instructions to authors
        (FILE) Editorial Board addresses
        (DIR)  Contents  of  forthcoming  issues
        (DIR) Contents of published issues
        (DIR) Subscriber Services (abstracts, papers, search)

                                Lynx Printing Options
```

If you want to save any file, type 'p' for 'print' and then select delivery to either a file on your own computer system, or as an electronic mail message.

```
                        PRINTING OPTIONS

   There are 43 lines, or approximately 1 page, to print.
   You have the following print choices
   please select one:

Save to a local file

Mail the file to yourself

Print to the screen
```

Then leave the system by quitting the browser.

Literature Exercise 6. USE WAIS TO CONDUCT A FULL-TEXT SEARCH OF LITERATURE REVIEWS

Task You are a Physics graduate student who needs to find some literature reviews in the field of fractofusion, to gain familiarity with current research in that field.

Approach Performing a keyword search in a broad database is an effective strategy to start finding literature on a topic. WAIS is likely to retrieve at least some documents because it runs the keyword search over the full-text of hundreds of sources.

Connect to a Public WAIS client

```
%telnet  quake.think.com
Trying 192.31.181.1...
Connected to quake.think.com.
Escape character is '^]'.
```

Note that this client is using simple WAIS, or SWAIS.

```
login: wais
Welcome to swais.
```

Give your email address as an identifier.

```
Please type user identifier (optional, i.e user@host):
clementg@solix.fiu.edu
```

Specify terminal emulation (commonly vt100).

```
TERM = (vt100) vt100
Starting swais (this may take a little while)...
This is the new experimental "wais" login on Quake.Think.COM
```

The opening screen explains how this public WAIS site is set up, with some basic instructions for starting a search.

```
As the total number of sources has passed the 500 mark, we've
found it's become virtually impossible to find a source from
the 25 screens of sources.
```

As indicated in the opening screen, this WAIS client automatically loads just one source for searching, the "Directory of Servers". This source is a catalog of all archives, databases etc. that are searchable by WAIS, with a description of what each source contains.

```
I have decided that instead of presenting you with all the
sources, I'll just give you the Directory of Servers as a
starting point.  To find additional sources, just select the
directory-of-server.src source, and ask it a question.  If you
know the name of the source you want, use it for the keywords,
and you should get that source as one of the results.  If you
don't know what source you want, then just ask a question that
has something to do with what you're looking for, and see what
you get.

Once you have a list of results, you should "u"se the result
you desire.You can "v"iew a result before you "u"se it, paying
close attention to the "description".
```

Select the source "directory-of-servers" by highlighting it and pressing the <spacebar>. Once selected, it is asterisked.

```
SWAIS           Source Selection                Sources:  1

 #    Server              Source              Cost

001: *[quake.think.com]  directory-of-servers    Free
```

You may start a search by pressing 'w' -- the system then prompts for keywords.

WAIS returned one source 'Cold-Fusion', which must have the search term 'fractofusion' in either its title or description.

To see the description of this source, highlight the item and press <return>.

WAIS retrieved this document because the search term 'fractofusion' occurs in the list of subjects covered.

Based on the description of the database, the user may decide if this is source is worthwhile.

Type 'q' to quit reading the display

To add the source 'Cold-fusion' to the active list of sources to be searched in WAIS, highlight it and press 'u' for 'use'.

Then type 's' to view the list of sources.

Keywords: **fractofusion**

Enter keywords with spaces between them; <return> to search; ^C to cancel

SWAIS	Search Results		Items: 1
#	Score	Source	Title	Lines

001: [1000]   (directory-of-se) cold-fusion   21

<space> selects, arrows move, w for keywords, s for sources, ? for help

---

SWAIS   Document Display
Page: 1
(:source
  :version 3
  :ip-address "152.2.22.81"
  :ip-name "SunSite.unc.edu"
  :tcp-port 210
  :database-name "cold-fusion"
  :cost 0.00
  :cost-unit :free
  :maintainer "cfh@sunsite.unc.edu"
  :subjects "cold fusion palladium titanium hydrogen deuterium
      nuclear energy CNF electrolysis metal hydrides
      fractofusion Pons Fleischmann Britz"
  :description "This is an annotated bibliography of published
      materials related to 'Cold Fusion' (the 'Pons &
      Fleischmann effect'). The material was selected and
      annotated by Dieter Britz of Aarhus Univ.,
      Denmark,britz@kemi.aau.dk. The bibliography emphasizes
      peer-reviewed journal papers. Updates appear in
      newsgroup sci.physics.fusion.
  .
  .   <display deleted>
  .

SWAIS   Search Results   Items: 1

  #   Score   Source   Title   Lines

001: [1000]   (directory-of-se)   cold-fusion   21

<space> selects, arrows move, w for keywords, s for sources, ? for help **s**

Note that there are now two sources. To restrict future searches to the Cold-fusion database (instead of the catalog of 500+ WAIS sources), make sure to select it with the **\<spacebar\>**, and de-select the Directory-of-Servers with the **\<spacebar\>**.

Running a keyword search now retrieves 40 documents in the Cold-fusion database. Each document contains at least one occurrence of the term 'fractofusion'.

To view a document, highlight it and press **\<return\>**

After viewing the selected document, press**\<return\>** to return to the list of 40 documents and read more; return to the list of sources with an '**s**'; or quit WAIS with a '**q**'.

```
SWAIS          Source Selection              Sources:  2

#     Server                        Source            Cost

001:  * [SunSite.unc.edu]    cold-fusion              Free
002:    [quake.think.com]    directory-of-servers     Free



Keywords: fractofusion

<space> selects, w for keywords, arrows move, <return>
searches, q quits, or ?


SWAIS          Search Results                   Items: 40

  #    Score Source            Title              Lines
001:  [1000] (cold-fusion)    Takeda T, Takizuka T;  10
002:  [942] (cold-fusion)     Yasui K;               19
003:  [412] (cold-fusion)     Davies JD, Cohen JS; 16
 .
 .     <display deleted>
 .


SWAIS                          Document Display
Page:  1
-------------------------
Davies JD, Cohen JS;
Ettore Majorana Int. Sci. Ser.: Phys. Sci. 1990, 52(Electromag.
Cascade Chem.Exot. At), 269.
"More on the cold fusion family".
** A theoretical physicists' view of cold fusion, in 1989. All
possibilities are critically examined, such as barrier
penetration, branching ratios, muon catalysis via cosmic
influx, and the micro-hot fractofusion. Some penetrating
comments are made. At the low energies of alleged cold fusion,
 .
 .     <text deleted>
 .

 shown to be unlikely. Experiments with tritiated water would
be most fruitful if fractofusion is the answer but the authors
warn of the dangers of T2 and especially T2O.


Press a key to continue <return>

<space> selects, arrows move, w for keywords, s for sources, ?
for help
```

Module 6: Electronic Publishing on the Internet

IN THIS MODULE:

Introduction
What's available
Quality and value
Access methods
Finding electronic publications in your field
How to cite electronic publications
Exercises

1. Use gopher or FTP to retrieve multiple files
2. Retrieve, download, and process compressed files
3. Make your preprint available as an e-print
4. Use a gopher directory to find an electronic journal
5. Browse an electronic journal on the Web and download
6. Retrieve and download an article in an unfamiliar format
7. View a WWW document in its HTML source file

Introduction

The printed scientific journal has been the primary medium for publishing scientific papers. But a combination of factors has raised the demand for a more economical, more efficient, quicker, and more flexible means of delivering new ideas and data to the research community. Soaring prices and shrinking budgets have caused large-scale cancellation of print subscriptions. The explosive growth in scientific information production has resulted in longer turnaround times from submission to production. And the changing communication needs of certain researchers, increasingly reliant on large datasets, complex processing methods, and models expressed in three or four dimensions, may no longer be accommodated in the printed page.

The Internet offers a promising alternative to print publishing. As a ubiquitous means of communication and resource sharing throughout the scientific community, it already delivers text, images, and data to desktops around the world. It stores and instantly transmits information in a multitude of formats, without size limitations. Moreover, its emergent capabilities for online multimedia presentation, for hyperlinking related information, and for searching, processing, and displaying selected information while online enables scientists to present and retrieve ideas and data in entirely new ways.

What's Available

Parallel Publishing

Science publishing on the Internet takes a variety of forms. On the conservative end of the spectrum are parallel publications—coexisting print and electronic journals providing the same information in differing formats. In some cases, the electronic version may offer enhanced functionality—continuous updating, hyperlinked references, and so on—but it still contains the same content as its printed counterpart. The Florida Entomological Society's *Florida Entomologist Online* , the electronic equivalent of the printed *Florida Entomologist*, is an example of a parallel publication now available on the Internet.[1]

Because publication of two discrete journals only adds to the total cost of production and delivery, parallel publishing is generally considered to be an experimental activity. It may, however, be an important first step leading to electronic-only publication.

Supplementary Material

Another scientific publishing trend increasing on the Internet is the online delivery of supplementary material from print journals. Figures, tables, data, software that can not be accommodated in a printed article due to limitations of size or format are made available on the Internet, with appropriate pointers included in the original article.

One of the most notable examples of supplementary material published on the Internet is the databank for the *Journal of Fluids Engineering*, established for the systematic delivery and archiving of research data, video clips, and other files supplementing the printed articles. The databank is maintained by the Scholarly Communications Project at Virginia Polytechnic Institute (VPI).[2]

Publication of supplementary material on the Internet may eventually lead to wholesale electronic publishing, as users now enjoying convenient and timely access to tables and figures may come to demand desktop access to the entire article.

Grey Literature Sources

Technical or scientific publications with limited distribution gain increased access when made available to the worldwide scientific community via the Internet. Technical report servers, for example, are increasingly used by government agencies, research agencies, and academic institutions to provide faster and wider access to documents that are otherwise difficult to find or obtain.

One of the most ambitious of the technical report services is the WATERS (Wide Area Technical Report Service) project, which offers a single point of access to dispersed technical report collections in computer science. Using a WAIS search, users of WATERS may search the abstracts or full text of technical reports, quickly retrieving and saving those of interest. According to its project objectives, WATERS aims to "speed up and increase the sharing of information in the computer science field and to encourage technology transfer."[3]

Network Science Journals

Network science journals may represent the most visionary development in Internet publishing to date. They may not 'look' like journals, taking the form of downloadable files of text and images; full-color pages with hyperlinks to related sources; and so on. Their units of access may be articles, not issues. But in each case, they provide high-quality research papers available exclusively on the Internet.

A number of new and traditional publishers have created science journals for the Internet. Among them are academic departments, research institutions, groups of colleagues, and scientific societies. Many of these journals are freely available in the public domain, including *Journal of Artificial Intelligence Research, Electronic Journal of Combinatorics, New York Journal of Mathematics*, and *Flora Online*.

Other journals on the Internet are available only through subscription for a fee. Not surprisingly, they represent some of 'fullest-featured' publications, with hyperlinked multimedia, advanced search options, and flexible formats for display and output.

Many of these journals represent joint ventures between traditional publishers and an established online information provider. Most notable is the increasing lineup of journals produced in partnership with the Online Computer Library Center (OCLC), an established information producer providing online services to libraries for many years.[4] For these journals, a scientific society or publisher provides editorial content and OCLC provides electronic publishing and delivery capabilities, including their proprietary user interface. Science journals produced by these partnerships include the Institution of Electrical Engineer's *Electronics Letters Online*, Elsevier's *Immunology Today Online*, the American Institute of Physics' *Applied Physics Letters Online*, and others.

For a more detailed discussion of network science journals, refer to the author's article "Evolution of a Species: Science Journals published on the Internet."[5]

Electronic Preprints (E-prints)

Electronic preprints, or e-prints, represent another popular form of scientific publishing emerging on the Internet. Authors submit electronic preprints, or e-prints to a common Internet node for widespread distribution in advance of traditional publication in print.

E-print services first developed in the physics community to enable researchers to leapfrog the peer-review process in order to get timely information out to their colleagues. According to Paul Ginsparg, maintainer of the Los Alamos Physics E-Print Archive, some e-prints have received several hundred requests the first day after submission, and as many as a thousand requests in subsequent weeks. Ginsparg has also observed that the papers continue to be retrieved even after their eventual publication in a journal, demonstrating that "people find the electronic format an easier means of access than physical access to a library."[6] The more recent move to initiate peer review for e-prints in physics increases the possibility that e-prints may eventually supplant traditional print publishing in this field.[7]

Quality and Value

Printed literature has three essential characteristics that ensure its authority and usefulness: rigorous review and editorial quality control; permanent archiving; and universal access. To merit equal consideration by the research community, scientific publications on the Internet must meet these same criteria.

Most scientific papers published on the Internet undergo the same editorial and peer-review process applied for their printed counterparts, ensuring their authority and reliability. Therefore, the Internet has no adverse effect on quality and value. It does, however have numerous advantages: quicker submission and review of manuscripts, enhanced functionality, more convenient and timely access, and possible cost-savings for publishers.

It is only the electronic preprint servers that have foregone the process of peer review and editorial control in favor of more timely delivery to the research community (and, as pointed out above, some are moving toward peer review as well). The effect of this trade-off can be measured only by those who use the literature. The increasing trend toward e-print publishing in some fields, and its lack of progress in others, reflects the differing assessment of preprint quality within the scientific community.

The permanence or stability of electronic archives on the Internet is more problematic. While fee-based journals are (not surprisingly) typically well-archived, those available in the public domain are not. In spite of well-intentioned publishers or other organizations committed to providing universal access to electronic journals, users find that archives are not always updated quickly, or backfiles are not loaded due to space constraints. These shortcomings serve to frustrate readers and discourage potential contributors.

Universal access to scientific publications on the Internet has also been slow to develop, as traditional indexing and abstract services are hesitant to cover any unproven or experimental journal. A few signs of a changing trend are evident, however. Such well established indexing services as *Index Medicus, Biological Abstracts, Computer and Control Abstracts*, and others are already selectively indexing a few network science journals. But journal publishers have a long way to go in ensuring widespread access to their electronic titles.

 21

Access Methods

As electronic publishers experiment with a variety of formats for presentation, delivery and storage of scientific publications on the Internet, users must contend with varying access methods and file formats. The most common means of access for scientific papers on the Internet are electronic mail (using MIME), file transfer protocol, Gopher, and the World-Wide Web. Each system is fully capable of storing and transmitting the highly visual forms of data and text so important to scientific information.

The format of the articles retrieved on the Internet also varies. With electronic mail, FTP and Gopher, for example, the text of the article may be available in one file, formatted as ascii, postscript, or TeX (look for extensions '.txt', '.ps' or 'tex', respectively). Corresponding graphics and other supplementary may be stored as separate files in image or data formats (*e.g.*, '.gif', '.pict'). Alternatively, the entire article may be available as a single text-plus-graphics file in a format such as Encapsulated Postscript ('file.eps') or Portable Document Format ('file.pdf') .

Files in any of the above formats may also be subsequently processed (compressed, archived, or both) to save storage space, transmission time, and ensure file integrity. For more information, see the Fact Sheet 'File Formats' in Section B of this guide.

To use publications available with any of the above methods, the researcher must retrieve and download the appropriate files, uncompress or unpack them if necessary, and open them with the appropriate software application and/or print them on a local printer.

In contrast to FTP- and Gopher-accessible files, publications on the World-Wide Web may be used "online." With graphics-capable client software such as Mosaic or NetScape and their associated "viewer applications," readers may search and browse the text, view images, run video clips, etc. Downloading or printing are secondary options.

The biggest drawback of scientific publishing on the World-Wide Web is the limitation of the current HTML specification in handling equations and other notations. At present, these elements of the publication must be included in the HTML source document as image files (in .gif or .xbm formats) that don't always align perfectly with the corresponding text.

However, several developments are underway to improve WWW publishing of scientific information. Enhancements to the HTML specification (termed "HTML +" or "HTML 3") accommodate tables, stand-alone equations, and other forms of scientific expression. For more information on HTML, in its current and future forms, refer to the HTML Fact Sheet in Section B of this guide.

Efforts to convert scientific documents prepared in LaTex format to HTML format, and to add hypertext capability to TeX documents, are also meeting success among certain scientific publishers. For further information about these developments, see Michael Grant's "Thoughts on Scientific HTML Documents"

> `http://www-isl.stanford.edu/~mcgrant/equations/`

and Paul Ginsburg's "HyperTex Page" maintained at the Los Alamos National Lab's Physics preprints archive

> `http://xxx.lanl.gov/hypertex`

or   `ftp://xxx.lanl.gov/pub/hypertex`

Finally, some scientific papers on the may be accessible using WAIS, a tool capable of performing full-text searches of textual documents and textual descriptions of graphics files. For searching ambiguous or high-frequency words in long documents, use a version of WAIS that has Boolean search capabilities (*e.g.*, IU BioWAIS or freeWAIS). Non-Boolean versions of WAIS (*e.g.*, simple WAIS, or SWAIS), which allow you only a single search term or word, are certain to retrieve numerous nonrelevant articles.

How to Find Electronic Publications in Your Subject

The strong interest in usage of electronic journals has led to the development of several useful resources for locating them online.

The best place to start is the annual *Directory of Electronic Journals, Newsletters, and Academic Discussion Lists.* Compiled and maintained by the Association of Research Libraries, the directory is available in searchable format on the Association's Gopher

```
gopher://arl.cni.org/scomm/11/edir
```

To receive announcements of new journals in planning or production, subscribe to the mailserver list: NewJour-L. Send email to:

```
listserv@e-math.ams.org
```

with the message

```
subscribe newjour-1 Yourfirstname Yourlastname
```

This list keeps users current between editions of the Association's e-journal directory mentioned above.

The CIC[8] Electronic Journals collection

```
gopher://gopher.cic.net:2000/11/e-serials/managed
```

established as a collaborative effort of librarians from CIC institutions, provides "a professionally managed collection of electronic journals of scholarly importance." This new service began in early 1995 and already offers links to many of the public-domain electronic journals, newsletters, and other serials available on the Internet.

For additional strategies and tools for finding Internet resources in your subject area, including electronic publications, refer to Section C of this guide.

How to cite electronic publications

For some guidelines, refer to the following publications:

Electronic Style: A Guide to Citing Electronic Information
Xia Li and Nancy B. Crane
Westport, CT: Meckler, 1993
ISBN: 0-88736-909-X

The Chicago Manual of Style, 14th ed.
(Includes a section on electronic publications)
Chicago: University of Chicago Press, 1993
ISBN: 0-226-10389-7

Also, many network science journals provide instructions on citing articles from that publication—look for such a file on the network server where you find the journal.

Endnotes

1. Archives of the *Florida Entomologist Online*, as well as information about this publishing project, are available from the Gopher of the Florida Center for Library Automation

    `gopher://sally.fcla.ufl.edu`

2. You may access the *Journal of Fluids Engineering* databank at

    `http://borg.lib.vt.edu/ejournals/JFE/jfe.html.`

3. WATERS is the joint project of Old Dominion University, SUNY Buffalo, University of Virginia, and Virginia Polytechnic Institute. To learn more about the project, or to use it to search for technical reports, point your Web browser to:

    `http://www.cs.odu.edu/WATERS/WATERS-GS.html`

4. For more information about OCLC's electronic journals, visit their Web server

    `http://www.oclc.org/`

5. Clement, Gail P., "Evolution of a Species: Science Journals Published on the Internet." *Database* 17, No. 5 (October 1994): 44-54.

6. Quote excerpted from a posting on August 25, 1994 to the electronic discussion group

    `VPIEJ-L@VTVM1.CC.VT.EDU`

7. Taubes, Gary, "Peer Review in Cyberspace." *Science*, vol. 266, no. 5187 (November 11, 1994): 967.

8. The Committee on Institutional Cooperation, or CIC, is a consortium of major Midwestern research institutions. CICNet is a midlevel network providing Internet services and other information resources to the region.

Publishing Exercise 1. USE GOPHER OR FTP TO RETRIEVE MULTIPLE FILES

Task You just read Izatt's article in *Chemical Reviews* (v. 94, p. 467) and saw that some of the tables and references are available only as supplementary files on the American Chemical Society (ACS) server, accessible by Gopher or FTP.

Approach: **(a)** Use Gopher to browse available files and save them one-by-one
(b) Use FTP to retrieve all desired files with one **mget** command

(a) If you prefer browsing to find what you want, try Gopher first.

Connect to the ACS Gopher

Select the appropriate directories.

Always a good idea to read the README first.

A choice of file formats is available. Shown here is that part of the README file displaying the filenames of the Postscript (.ps) files.

To save these .ps files, get out of the README file by typing 'u'; then move the arrow to the desired file and type 's' to save to your own computer. You will have to repeat this save step for each file desired.

Type 'q' to quit Gopher and return to your own system prompt. From there you may download the files and send them directly to a laser printer, or view them onscreen with a postscript viewer such as Ghostview.

```
% gopher  acsinfo.acs.org

--> ACS Publications/
      --> Supplementary Material/
            --> ChemicalReviews/
                  --> v094/
                        -->p467/
                              -->README
```

These files are from the supplementary material provided with the Reed M. Izatt paper (Chemical Reviews, vol. 94, page 467). The information includes high-temperature thermodynamic data for the interaction of protons and metal ions with inorganic and organic ligands (Tables 1 and 2), a listing of the equations used to derive the data (Table 3), and a complete list of references.

The files are available in three formats: MicroSoft Word (.MSW), RTF (.RTF) and PostScript (.ps). The PostScript and RTF files can be transmitted by either gopher or FTP.

.
.   <text deleted>
.

PostScript Format

Base File	Ext.	Byte Cnt.	Trans. Mode	Printed Pgs. in MS. Word
Refs	.ps	1264355	ASCII	32 p.
Table_I.1	.ps	757760	ASCII	116 p.
Table_I.2	.ps	1428603	ASCII	154 p.
Table_I.3	.ps	1555516	ASCII	146 p.
Table_I.4	.ps	1414377	ASCII	92 p.
Table_II	.ps	636398	ASCII	25 p.
Table_III	.ps	1042489	ASCII	59 p.

(b) Connect to the ACS server using FTP so that you can retrieve all desired files at once with the **mget** command.

Connect to the ACS anonymous FTP server.	`% ftp acsinfo.acs.org`
	`Connected to acsinfo.acs.org.`
	`220 acsinfo FTP server (SunOS 4.1) ready.`

Connect to the ACS anonymous FTP server.

```
% ftp acsinfo.acs.org
Connected to acsinfo.acs.org.
220 acsinfo FTP server (SunOS 4.1) ready.
```

Login as an anonymous user

```
Name (acsinfo.acs.org:clementg): anonymous
331 Guest login ok, send ident as password.
```

Enter email address for a password (good netiquette!)

```
Password: clementg@solix.fiu.edu
230 Guest login ok, access restrictions apply.
```

Display current directory to see where you are

```
ftp> dir
150 ASCII data connection for /bin/ls (131.94.64.200,3297) (0
bytes).
total 18
dr-xr-xr-x  2 2000     102       512 Nov 23  1993 .cap
```

Here's the subdirectory for *Chemical Reviews*

```
drwxrwxr-x  3 2000     102       512 Mar 25 20:05 ChemicalReviews
dr-xr-xr-x  6 2000     102       512 Feb 14 19:53 JACS
drwxrwxr-x  3 2000     102       512 Apr  1 13:47 Publications
.
.      <display deleted>
.
```

Change to that subdirectory

See what files are available

```
ftp> cd ChemicalReviews
250 CWD command successful.
ftp> dir
150 ASCII data connection for /bin/ls (131.94.64.200,3298)
total 1
drwxrwxr-x  3 2000     102      1024 Apr  5 19:58 v094
```

Change to the subdirectory for the pertinent volume

```
ftp> cd v094
250 CWD command successful.
ftp> dir
150 ASCII data connection for /bin/ls (131.94.64.200,3299)
total 2
-rw-rw-r--  1 103      102       135 Mar 28 16:19 README
drwxr-xr-x  2 103      102      1024 Apr  5 20:03 p467
```

Change to the subdirectory for the pertinent article

All the desired files are here, in postscript (.ps) format

```
ftp> cd p467
250 CWD command successful.
ftp> dir
150 ASCII data connection for /bin/ls (131.94.64.200,3300)
total 10095
-rw-r--r--  1 103      102         2892 Apr 12 12:06 README
-rw-rw-rw-  1 103      102        81920 Apr  5 20:03 Refs.MSW
-r--r--r--  1 103      102       106646 Mar 25 20:41 Refs.RTF
-r--r--r--  1 103      102      1264355 Mar 25 20:51 Refs.ps
.
.      <display deleted>
.
```

```
-rw-rw-rw-  1 103        102      85504 Apr  5 20:02 Table_III.MSW
-r--r--r--  1 103        102     157044 Mar 25 20:42 Table_III.RTF
-r--r--r--  1 103        102    1042489 Mar 25 20:52 Table_III.ps
```

Toggle interactive prompting off so that system does not prompt in between each file transfer

```
ftp> prompt
Interactive mode off.
```

Issue multiple get (**mget**) command with the wildcard character to get all files ending with '.ps'

Note: You do not need to issue a command to switch transfer mode from ASCII to BINARY because postscript files are generally transferred as the former.

System confirms the success of each file transfer after completion.

```
ftp> mget *.ps
150 ASCII data connection for Refs.ps (131.94.64.200,3302)
(1264355 bytes).
226 ASCII Transfer complete.
local: Refs.ps remote: Refs.ps
1281526 bytes received in 32 seconds (39 Kbytes/s)
150 ASCII data connection for Table_I.1.ps (131.94.64.200,3304)
(757760 bytes).
226 ASCII Transfer complete.
local: Table_I.1.ps remote: Table_I.1.ps
787754 bytes received in 17 seconds (46 Kbytes/s)
   .
   .     <display deleted>
   .

150 ASCII data connection for Table_III.ps (131.94.64.200,3320)
(1042489 bytes).
226 ASCII Transfer complete.
local: Table_III.ps remote: Table_III.ps
1084357 bytes received in 32 seconds (33 Kbytes/s)
```

Quit FTP

```
ftp> quit
221 Goodbye .
```

As with the Gopher exercise above, you may download the files and send them directly to a laser printer, or view them onscreen with a postscript viewer such as Ghostview.

Publishing Exercise 2. RETRIEVE, DOWNLOAD, AND PROCESS A SET OF COMPRESSED/ENCODED FILES

Task You are interested in seeing the latest work in the area of chiral perturbation theory, a research front in the field of high energy physics.

Approach Over the last few years, physics researchers have gravitated toward e-print servers to publish their latest research results. The largest collection of physics e-prints is maintained by Los Alamos National Laboratories. Accessing the archives by Gopher will allow for easy browsing and retrieval.

Connect to the Gopher at Los Alamos.	`% gopher mentor.lanl.gov`
	`--> High Energy Physics -- Lattice/`
Choose menu entries as appropriate.	`--> Index to Titles, Authors, and Abstracts hep-lat <?>`

```
+---------Index to Titles, Authors and Abstracts hep-lat-----+
| Words to search for                                        |
|                                                            |
|   chiral perturbation                                      |
|                                                            |
|                [Cancel: ^G] [Erase: ^U] [Accept: Enter]    |
+------------------------------------------------------------+
```

Enter keywords in the search box.

Gopher returns a menu of 'hits' matching one or both keywords.

Select the desired preprint.

```
Index to Titles, Authors and Abstracts: chiral perturbation

-->1.9411005_hep-lat_Chiral Perturbation Theory and Quench ../
    2.9206005_hep-lat_MASSLESS DECOUPLED DOUBLERS: CHIRAL  ../
    3.9312067_hep-lat_QUENCHED CHIRAL PERTURBATION THEORY  ../
        .
        .           <display deleted>
        .
```

For this preprint, both a text abstract and the paper itself are available.

If desired, read the abstract first.

```
--> 1.  Abstract:9411005_ Chiral Perturbation Theory and...
    2.  Paper (Combined Source).


Abstract:9411005_ Chiral Perturbation Theory and the Quenched
+-----------------------------------------------------------+
Paper: hep-lat/9411005
From: Maarten Golterman <maarten@aapje.wustl.edu>
Date: Thu, 3 Nov 94 17:30:04 -0600

Title: Chiral Perturbation Theory and the Quenched
Approximation of QCD
Author: Maarten F.L. Golterman
```

Type ' **u** ' to quit browsing the abstract.

```
Comments: 32 pages, uses jnl macropackage, two figures missing,
hardcopy available on request
Report-no: Wash. U. HEP/94-63
\\
Lectures given at the XXXIV Cracow School of Theoretical
Physics, Zakopane, Poland.
The quenched approximation for QCD is, at present and in the
foreseeable future, unavoidable in lattice calculations with
.
.        <display deleted>
.
```

From the menu, select the entry for the 'Paper'.

```
        1.  Abstract:9411005_ Chiral Perturbation Theory and...
-->    2.  Paper (Combined Source).
```

The file is not simple text (ASCII), so it will not display in full view on the screen.

But the header describes what type of file it is and how to process it once retrieved.

As noted, the file is a set of related files (probably the manuscript and some graphics) that have been archived into a single file (tar), compressed or zipped to save room (Z), and encoded (uu) for safe transmission over the network.

```
Paper (Combined Source) (84k)
+-------------------------------------------------------------+
%-Eprint Note: Typing File etc/eprints/hep-
lat/papers/9411/9411005
%Paper: hep-lat/9411005
%From: Maarten Golterman <maarten@aapje.wustl.edu>
%Date: Thu, 3 Nov 94 17:30:04 -0600

#!/bin/csh -f
# Note: this uuencoded compressed tar file created by csh
script  uufiles
# if you are on a unix machine this file will unpack itself:
# just strip off any mail header and call resulting file, e.g.,
qchpt.uu
# (uudecode will ignore these header lines and search for the
begin line below)
# then say          csh qchpt.uu
# if you are not on a unix machine, you should explicitly
execute the commands:
#    uudecode qchpt.uu;    uncompress qchpt.tar.Z;    tar -xvf
qchpt.tar
+-------------------------------------------------------------+
[PageDown: <SPACE>] [Help: ?] [Exit: u]

u
```

To retrieve the file, quit browsing with the '**u**' and get back to the menu.

Move the arrow to the entry for the Paper and press '**s**' to save as a file on your own computer.

Then quit Gopher with a '**q**'.

```
        1.  Abstract:9411005_ Chiral Perturbation Theory and ...
-->    2.  Paper (Combined Source).

     +------------------------Paper (Combined Source)-----------+
     |                                                          |
     | Save in file:                                            |
     |                                                          |
     | qchpt.uu                                                 |
     |                       [Cancel: ^G] [Erase: ^U] [Accept: Enter] |
     +----------------------------------------------------------+
```

Back at your system prompt, you will see the file has been saved.

As the header of the pre-print indicated, the file may be processed step-by-step as follows:

Decode (creates a compressed file)

Uncompress (creates an archived file)

Unpack the archive file, and out spill all the constituent files.

(NOTE: For further information about various file formats on the Internet and how to deal with them, see the "File Formats" Fact Sheet in the next section of this guide.)

```
% dir
-rw-rw-rw-  1 clement       86764 Nov 11 16:47 qchpt.uu
```

```
% uudecode qchpt.uu  (creates  a  file  qchpt.tar.Z)
```

```
% uncompress qchpt.tar.Z (creates  a  file  qchpt.tar)
```

```
% tar  -xvf  qchpt.tar
zakopane.tex
fig1.ps
fig2.ps
fig3.ps
```

Publishing Exercise 3. MAKE YOUR PREPRINT AVAILABLE AS AN E-PRINT

Task You are a computer scientist and just finished a paper on neural networks. You would like to post it to colleagues for review and comment. You know that E-prints on this topic are archived at a site called 'neuroprose', but don't know what kind of server this site uses, where it is on the Internet, or what the requirments are for adding a preprint to the collection.

Approach Most E-print collections are stored, at a minimum, as FTP archives (Gopher, WAIS and WWW may sometimes provide alternate means of access.) All files and directories on anonymous FTP servers are searchable with Archie, as long as their names are known. In this case, you presuppose the name 'neuroprose'.

After getting the address of the E-print server from Archie, you can then connect to it and look for an 'Instructions to Contributors' file.

Connect with one of the many public Archie sites.

```
%telnet  archie.rutgers.edu
login: archie
# Terminal type set to `vt100 24 80'.
# `erase' character is `^?'.
# `search' (type string) has the value `sub'.
```

The default search type 'sub' means the match may not be an exact string but rather, a substring of a longer term.

```
archie>  set  mailto  clementg@solix.fiu.edu
archie>  prog  neuroprose
```

Enter address where results should be mailed.

Archie notifies you of the approximate time to complete the search.

```
# Search type: sub.
# Your queue position: 3
# Estimated time for completion: 3 minutes, 7 seconds.
working...  =O=O=O=O=O=O=O=O=O=O=O=O
```

Search results are displayed in one continuous scroll.

Here is a 'neuroprose' directory at the FTP server at Ohio State. Its recent update suggests that this is an active directory.

```
Host ftp.cis.ohio-state.edu    (128.146.8.52)
Last updated 08:11  8 Nov 1994

Location: /
DIRECTORY  drwxr-xr-x 19456 bytes  07:00 1 Nov 1994  neuroprose

Location: /neuroprose
FILE    -rwxrwxrwx     10 bytes  18:00 18 Apr 1994  neuroprose
```

Any other directory or file name containing the substring 'neuroprose' is also retrieved.

```
Host cs.dal.ca    (129.173.4.5)
Last updated 05:12 14 Oct 1994

Location: /comp.archives/comp.ai.neural-nets
FILE    -r--r--r--   2554 bytes  10:16 13 Jun 1994  tdrbf-
```

You may request a copy of the search results sent to your own electronic mail box.

If you mailed the results of the Archie search, you will receive a mail message from Archie, containing the same results that were displayed at the end of the search.

Now you are ready to FTP directly to the site reported by Archie.

Login as anonymous; use email address for the password.

Change to the /pub directory; and then to /neuroprose.

Always good to read the README first!

The 'more' command requests that the file be displayed on the screen, for browsing.

Eureka! Complete instructions for pre-print contributors are provided.

```
paper-available -in-neuroprose
.
.
.

archie> mail
archie> quit
# Bye.
Connection closed by foreign host.

From:   SMTP%"archie-errors@dorm.rutgers.edu" 13-NOV-1994
20:42:46.80

Subj:   archie [prog neuroprose] part 1 of 1

Host nervous.cis.ohio-state.edu    (164.107.143.6)
Last updated 13:17 11 Nov 1994


Location: /pub
FILE    -rwxrwxrwx       28 bytes  18:00 16 Jun 1993   neuroprose
.
.      <text deleted>
.


%ftp   ftp.cis.ohio-state.edu
Name (ftp.cis.ohio-state.edu:clementg): anonymous
331 Guest login ok, send your complete e-mail address as
password.
Password:clementg@solix.fiu.edu
230 Guest login ok, access restrictions apply.
ftp> cd pub
250 CWD command successful.
ftp> cd neuroprose
250-Please read the file README
250-  it was last modified on Wed Jul 20 09:43:24 1994 - 116
days ago
ftp> get README |more
200 PORT command successful.
150 Opening ASCII mode data connection for README (5564 bytes).
 Anonymous FTP on archive.cis.ohio-state.edu (128.146.8.52)
                pub/neuroprose directory

This directory contains technical reports as a public service
to the connectionist and neural network scientific community
which has an organized mailing list (for info: connectionists-
request@cs.cmu.edu)

Researchers may place electronic versions of their preprints in
this directory, announce availability, and other interested
researchers can rapidly retrieve and print the postscripts.
This saves copying, postage and handling, by having the
interested reader supply the paper.  We strongly discourage the
```

merger into the repository of existing bodies of work or the use of this medium as a vanity press for papers which are not of publication quality.
.
.   <text deleted>
.
--More--

Press 'q' to quit browsing

Transfers the README file to your own computer for future reference.

The destination filename will default to that of the source file.

Close FTP session.

```
ftp> get README
200 PORT command successful.
150 Opening ASCII mode data connection for README (5564 bytes).
226 Transfer complete.
local: README remote: README
5713 bytes received in 0.62 seconds (8.9 Kbytes/s)
ftp> quit
221 Goodbye.
```

Publishing Exercise 4. USE A SEARCHABLE GOPHER DIRECTORY TO FIND AN ELECTRONIC JOURNAL

Task   You just found out about a new refereed journal in your field, *Electronic Journal of Combinatorics*, published only in electronic format on the Internet. How do you find the journal?

Approach   Check the *Directory of Electronic Journals and Newsletters*, published by the Association of Research Libraries (ARL). It's available on the ARL Gopher.

Connect to ARL Gopher

As denoted by the <?>, the Directory is available as a searchable file

```
% gopher arl.cni.org

--> Scholarly Communication/
      --> Directory of Electronic Journals...../
           -->1994 ARL Directory of Electronic Journals.../
                --> Search Gopher Edition of Directory <?>


                     Internet Gopher Information Client v2.0.16
```

Selecting this option on the menu calls up a search box Since the search run by Gopher uses simple WAIS, boolean operators are not available, and multiple terms are assumed to be connected with an OR. Therefore, avoid the title words 'electronic' and 'journal', since all files containing either word would be retrieved. Choose instead the low-frequency word 'combinatorics'.

```
1994 ARL Directory of Electronic Journals and Newsletters

         1.  Introduction to Gopher Edition of the ARL Directory-
-->      2.  Search Gopher Edition of Directory <?>
         3.  Electronic Journals and 'Zines
         4.  Newsletters, Digests and Reference

+---------------Search Gopher Edition of Directory---------+
| Words to search for                                      |
|                                                          |
|   combinatorics                                          |
|                                                          |
| [Help: ^-]  [Cancel: ^G]                                 |
+----------------------------------------------------------+
```

Gopher then returns a list of matching items.

Selecting an item then brings up the description of the journal.

Two access methods are available: electronic mail and direct access on the WorldWide Web.

Type 'u' to quit browsing, and 'q' to leave Gopher.

```
-->  1.  Electronic Journals and 'Zines


------------------------------------------------------------
Electronic Journal of Combinatorics

A refereed journal of discrete mathematics which welcomes
submissions of papers in all branches of combinatorics, graph
theory, discrete algorithms, etc.

Message calkin@math.gatech.edu with your name and email address
http://ejc.math.gatech.edu:8080/Journal/journalhome.html

Herbert Wilf, Editor-in-Chief
+---------------------------------------------------------------+
```

Publishing Exercise 5. BROWSE THROUGH AN ELECTRONIC JOURNAL ON THE WORLD-WIDE WEB, AND TRY DOWNLOADING AN ARTICLE.

Task Connect to the *Electronic Journal of Combinatorics* on the World-Wide Web. Try to 'use' the journal online -- follow the links to retrieve and read something from the journal.

Approach If you have access to a Web browser on your computer, point directly to the URL:

```
%lynx  http://ejc.math.gatech.edu:8080/Journal/
                                     journalhome.html
```

If you do not have Web access directly, you may Telnet to a public site for Lynx, the text-based, fullscreen browser for the World-Wide Web.

```
%  telnet  kufacts.cc.ukans.edu
  login: kufacts
```

Press '**g**' to go to another site on the WorldWide Web.
At the command line, type in the URL

```
http://ejc.math.gatech.edu:8080/Journal/journalhome.html
```

Once you are at the journal site, the commands and options are the same regardless of which approach you follow.

Use the **<tab>** key to move from one hyperlink to the next.
Use the **<spacebar>** to move from one page to the next.

Note that many of the articles (formatted in TeX or PostScript) as well as images will not display on the screen. They are not formatted as simple text and thus need to be downloaded and viewed offline.

Try downloading an article or graphic by highlighting the given item and then pressing '**d**' to download. You will then be prompted for a filename.

Here is what the cover page of the journal looks like in text-only format with Lynx.

Cover page for the Electronic Journal of Combinatorics (p1 of 3)

[IMAGE]

WITH THE COOPERATION OF THE AMERICAN MATHEMATICAL SOCIETY

VOLUME 1

Articles

  * A1: Donald E. Knuth,
      + The Sandwich Theorem. (48pp)

Research Papers

  * R1: Dominique Foata and Doron Zeilberger,
      + Combinatorial Proofs of Capelli's and Turnbull's Identities
        from Classical Invariant Theory. (10pp)
  * R2: Richard A. Brualdi and Stephen Mellendorf,
      + Two Extremal Problems in Graph Theory. (10pp)
  * R3: Stan Wagon and Herbert S. Wilf,

-- press space for next page --
Arrow keys: Up and Down to move. Right to follow a link; Left to go back.
 H)elp O)ptions P)rint G)o M)ain screen Q)uit /=search [delete]=history list

Publishing Exercise 6. RETRIEVE AND DOWNLOAD AN ARTICLE IN A
FORMAT YOU DON T RECOGNIZE

Task You are a researcher who wants to read an article published in *Florida Entomologist*. You know it is available online at the Gopher of the Florida Center for Library Automation. But you don't know what format the file(s) will be in, and what method is needed to retrieve, process, and use the article.

Approach Find the journal's archive site at FCLA and see what format the article is in. It is then possible to determine the best means of retrieval, processing, etc.

Connect to the FCLA Gopher, and make the appropriate selection.

```
% gopher sally.fcla.ufl.edu

--> Florida Entomologist Online/


            Florida Entomologist Online

        1. About Florida Entomologist Online.
--> 2. Access to available volumes,issues .../
        3. Acrobat 2.0 reader for Windows <PC Bin>

        --> Vol. 77(3): 301-396 (Sept. 1994)/


            Vol. 77(3): 301-396 (Sept. 1994)

        1.  About the names of these PDF files.
        2.  FE77P301.PDF     Braman, S.K., et al.  <Bin>
        3.  FE77P305.PDF     Landolt, P.J.            <Bin>
        4.  FE77P313.PDF     Atkinson and Peck  <Bin>
        5.  FE77P330.PDF     Vega, R.E., et al.  <Bin>
--> 6.  FE77P334.PDF     Tingle, F.C., et al.   <Bin>

    +-------FE77P334.PDF     Tingle, F.C., et al. -----+
    |                                                  |
    |  Save in file:                                   |
    |                                                  |
    |  FE77P334.PDF-----Tingle,-F.C.,-et-al.-          |
    |                                                  |
    | [Cancel: ^G] [Erase: ^U] [Accept: Enter]         |
    +--------------------------------------------------+

15. FE77P387.PDF      In Memoriam, H.R. Gross <Bin>
.
.       <text deleted>
.


            Florida Entomologist Online
```

Eager to get started, you jump right to the journal issues.

When you get to the desired issue, you find that the articles are in 'pdf' format. Selecting them does not bring up the text of the articles. Instead, a save box automatically opens.

Save the file and then go back to the main menu to see if you can find out what a 'pdf' file is.

Maybe the 'About Florida Entomologist Online' file will help.

The 'About' file provides access information for the journal articles.

Now you know to download the files, along with the free viewer, and then view them on a PC or send them to a printer.

```
 --> 1.  About Florida Entomologist Online.
     2.  Access to available volumes, issues.../
     3.  Acrobat 2.0 reader for Windows <PC Bin>

About Florida Entomologist Online (1k)
81%
+---------------------------------------------------------+
About Florida Entomologist Online

Florida Entomologist Online (FEO) is published in
parallel with, and is identical to, the traditionally
published Florida Entomologist (An International
Journal for the Americas).  The articles in FEO
are .PDF files, a Portable Document Format, produced
by Adobe's Acrobat 2.0.  They can be viewed, searched,
and printed with a FREE Acrobat 2.0 Reader, which can
be downloaded from this menu.  If you print articles
from FEO on a good printer, you will have the
equivalent of a photocopy of the traditionally
published article.

FEO was undertaken to further this vision of the
future of primary scientific publication:

"Any scientist who is linked to the developing
worldwide electronic information network (presently
termed the Internet) will be able to view and to print
any article in any journal published by a scientific
society.  Printing from the network will yield
hardcopy equal to a photocopy or reprint of the
article."  [endorsed by the Executive Committee of the
Florida Entomological Society, 10 May...
.
.       <text deleted>
.


Florida Entomologist Online

     1.  About Florida Entomologist Online.
     2.  Access to available volumes, issues.../
 --> 3.  Acrobat 2.0 reader for Windows <PC Bin>


  +--------------Acrobat 2.0 reader for Windows-------+
  | Save in file:                                     |
  |                                                   |
  |  Acrobat-2.0-reader-for-Windows                   |
  |                                                   |
  |   [Cancel: ^G] [Erase: ^U] [Accept: Enter]        |
  +---------------------------------------------------+
```

Back on the Gopher menu, you may move to the option for the reader and then press 's' to save or 'D' to download.

In this case, you saved the reader to your shell account. You will then need to download it to your PC using your communications software.

Publishing Exercise 7. VIEW A WWW DOCUMENT IN ITS HTML SOURCE FILE.

Task You are interested in publishing an article in *Complexity International*, a scientific journal on the World-Wide Web. You wish to submit the article marked up in HTML format, and know you can figure out how to do it by finding a comparable Web document and studying its HTML source file.

Approach Retrieve and examine some comparable articles already published in *Complexity International*. Almost all Web clients allow you to view and/or save the unrendered source of a given document (that is, the original HTML files that are 'translated' by the Web client into a 'polished' journal article).

Point your Web browser to the Uniform Resource Locator (URL) for *Complexity International*; browse until you find an article of interest.

Shown here is the HTML source file for a section of A. Bouzerdoum's article "Convergence of Symmetric Shunting Competitive Neural Networks", *Complexity International*, vol. 1, (April 1994).

Most Web browsers provide the option to view and save the HTML source for the article. On Lynx, press the '\' (backslash) key; to save the file, press '**P**' to print and then choose the option for saving to a file.

In a graphical client such as MacWeb, select the '**Options**' menu to view the source file, and then the '**File**' menu to save it.

HTML source files are saved in ascii format.

```
%lynx http://life.anu.edu.au/ci/ci.html


<HEAD>
<TITLE> Boundedness of solutions</TITLE>
</HEAD>
<BODY><P>
 <HR> <A NAME=tex2html168 HREF=section3_3.html><iMG
SRC="http://life.anu.edu.au/ci/icons/next_motif.gif"></A> <A
NAME=tex2html166 HREF=section3_2.html><iMG
SRC="http://life.anu.edu.au/ci/icons/up_motif.gif"></A> <A
NAME=tex2html162 HREF=subsection3_2_2.html><iMG
SRC="http://life.anu.edu.au/ci/icons/previous_motif.gif"></A>
<BR>
<B> Next:</B> <A NAME=tex2html169 HREF=section3_3.html>
Convergence of solutions</A>
<B>Up:</B> <A NAME=tex2html167 HREF=section3_2.html> Bounded-
input bounded-output (BIBO) </A>
<B> Previous:</B> <A NAME=tex2html163 HREF=subsection3_2_2.html>
Negativity of solutions</A>
<HR> <P>
<H2><A NAME=SECTION0002300000000000000> Boundedness of
solutions</A></H2>
<P>
<P><A NAME=thbibo><iMG
SRC="http://life.anu.edu.au/ci/icons/invis_anchor.xbm"></A><iMG
ALIGN=BOTTOM ALT="" SRC="_628_theorem224.xbm"><P>
<i>Proof:</i> Since the activation functions
are continuous, the outputs of the dynamical systems (<A
HREF=section3_1.html#eqsinn>1</A>) are bounded if the
trajectories are bounded. To prove boundedness of trajectories
it suffices to show that the positive activities are
bounded from above while the negative ones are bounded from
below.
If an activity <iMG ALIGN=BOTTOM ALT=""
SRC="_628_tex2html_wrap1216.xbm"> is positive over an interval
of time
```

Compare this HTML source file with final page rendered by the Web client (shown on the next page).

Note that the mathematical formulae are presented as separate image files linked to the text.

```
<iMG ALIGN=BOTTOM ALT="" SRC="_628_tex2html_wrap1218.xbm">,
then:
<P><iMG ALIGN=BOTTOM ALT="" SRC="_628_displaymath234.xbm"><P>
Following the same steps taken in the proof of Lemma <A
HREF=subsection3_2_1.html#lmpbound>1</A>,
it can be readily shown that:
<P><iMG ALIGN=BOTTOM ALT="" SRC="_628_displaymath244.xbm"><P>
where as before <iMG ALIGN=BOTTOM ALT=""
SRC="_628_tex2html_wrap1220.xbm">.
<P>
Likewise if the activity <iMG ALIGN=BOTTOM ALT=""
SRC="_628_tex2html_wrap1222.xbm"> is negative over an interval
of time
<iMG ALIGN=BOTTOM ALT="" SRC="_628_tex2html_wrap1224.xbm">,
then:
<P><iMG ALIGN=BOTTOM ALT="" SRC="_628_displaymath260.xbm"><P>
This implies that:
<P><iMG ALIGN=BOTTOM ALT="" SRC="_628_displaymath269.xbm"><P>
where <iMG ALIGN=BOTTOM ALT=""
SRC="_628_tex2html_wrap1226.xbm">.
<P>
Since the above is true for all activities, <iMG ALIGN=BOTTOM
ALT="" SRC="_628_tex2html_wrap1228.xbm">
<iMG ALIGN=BOTTOM ALT="" SRC="_628_tex2html_wrap1230.xbm">, we
can conclude that the trajectories
of (<A HREF=section3_1.html#eqsinn>1</A>) are bounded if the
input patterns are bounded; consequently,
the shunting inhibitory neural networks (<A
HREF=section3_1.html#eqsinn>1</A>) are BIBO stable.
<P>
<P><HR>
</BODY>
<P><ADDRESS>
<a href="http://life.anu.edu.au/ci/vol1/ci1.html"><i>Complexity
International</I> (1994) 1</A>
```

Boundedness of solutions

Theorem 1 *If the activation functions, $f_j \ \forall j = 1, \ldots, n$, are continuous and nonnegative on the entire real axis, then the shunting recurrent neural networks (1) are bounded-input bounded-output stable dynamical systems.*

Proof: Since the activation functions are continuous, the outputs of the dynamical systems (1) are bounded if the trajectories are bounded. To prove boundedness of trajectories it suffices to show that the positive activities are bounded from above while the negative ones are bounded from below. If an activity $x_i(t)$ is positive over an interval of time $[t_0, t_0 + \xi]$, then:

$$\dot{x}_i(t) \leq L_i(t) - a_i x_i(t) \ \forall t \in [t_0, t_0 + \xi]$$

Following the same steps taken in the proof of Lemma 1, it can be readily shown that:

$$x_i(t) \leq \max(B_i, x_i(t_0)) \ \forall t \in [t_0, t_0 + \xi],$$

where as before $B_i = \sup\{L_i(t)/a_i \ \text{for} \ t \geq t_0\}$.

Likewise if the activity $x_i(t)$ is negative over an interval of time $[t_1, t_1 + \eta]$, then:

$$\dot{x}_i(t) \geq L_i(t) - a_i x_i(t) \ \forall t \in [t_1, t_1 + \eta]$$

This implies that:

$$x_i(t) \geq \min(b_i, x_i(t_1)) \ \forall t \in [t_1, t_1 + \eta],$$

where $b_i = \inf\{L_i(t)/a_i \ \text{for} \ t \geq t_1\}$.

Since the above is true for all activities, $x_i(t) \forall i = 1, \ldots, n$, we can conclude that the trajectories of (1) are bounded if the input patterns are bounded; consequently, the shunting inhibitory neural networks (1) are BIBO stable.

Module 7: Sharing Data and Resources

IN THIS MODULE:

Introduction

What's available

Sharing and accessing methods

Quality and value

How to Find data and software resources in your field

Exercises

1. Locate, retrieve and process software for Macintosh

2. Locate, retrieve and process software for the PC

3. Use NETLIB to find mathematical software

4. Download and view an image from the World-Wide Web

Introduction

Central to any scientific research endeavor is a core of primary resources: either quantitative data, used as a basis for analysis, correlation or interpretation; or software, algorithms or source code, used for manipulation, visualization, or modelling. Though generated to answer an immediate research question, these resources may be potentially useful to other researchers in the same or other disciplines, for related or separate studies. Consequently, funding agencies usually require grant recipients to share data and resources as a condition of their awards. Publishers are also beginning to require, before accepting a manuscript, contribution of corresponding data to public domain data archives. For these many reasons, it is common practice to make scientific data and related resources widely available to the global scientific community.

Nonetheless, researchers do not always find it easy to gain access to the primary data and programs generated by their colleagues. They may find it difficult to determine what resources are available. Once they identify desired data or programs, they may have difficulty procuring them. Often they must prevail upon the original investigator to deliver files on diskette or some other format. And tools and techniques developed by one group of scientists to meet their specific needs for data gathering, management, analysis, and sharing may not be available to others.

Problems of duplicated effort, lost opportunity, and other inefficient practices have prompted some researchers to centralize data resources for their area of specialty. Data from dispersed sources may be compiled into a large centralized databank service, from which investigators may retrieve (and often buy!) data on demand But the high costs, cumbersome access requirements, copyright and other legal problems surrounding these traditional databank services have often deterred scientists from using or contributing to them.

The Internet offers a more promising alternative for effective, economical, and universal access to scientific data and software. Network capabilities for storage, searching, retrieval and delivery facilitate rapid dissemination of data on a worldwide scale. Existing Internet protocols bring together the pieces of highly complex or distributed research into one system, providing users a single point of access. Popular tools such as Gopher and the World-Wide Web offer easy-to-use interfaces for browsing, selection, and even online display. As noted by biologist David Green in his article "Databasing diversity—A Distributed, Public-domain Approach,"[1] by providing a means of freely exchanging data and other resources on a global scale, the Internet has already changed the way in which research is done.

Additionally, new Internet-based data sharing applications are beginning to offer scientific communities new ways of sharing data an other primary resources. Emerging new tools such as Collage facilitate a range of collaborative activities among dispersed colleagues, including shared data analysis, manipulation, and visualization.[2]

Finally, the Internet is also changing how and what new data may be collected. Thanks to numerous initiatives on the local, national, and international scale, we are seeing the emergence of new tools and techniques for sharing or remotely controlling instruments, and recording or relaying data in real-time. Applications of these "state-of-the-art" Internet tools include the remotely-controlled electronic microscope at the University of California, San Diego, enabling scientists to position the viewing stage from their own network-connected workstations[3]; the Hubble Space Telescope, which recorded and immediately transmitted to Earth early images of the collision between comet Shoemaker-Levy and planet Jupiter; and Woods Hole Oceanographic Institution's Jason Remotely Operated Vehicle (ROV), which instantly transmits data from the seafloor to modelers (as well as hundreds of students) on shore.[4]

As these emergent tools develop and mature, they are certain to become essential tools for cooperation and collaboration throughout the scientific community. And like the earlier Internet tools, originally developed to facilitate scientific communications and resource sharing (*e.g.*, email, FTP, and, more recently, WWW), these new emergent tools are likely to become essential resources for any Internet user, ultimately benefiting all sectors of the community, worldwide.

22

What's Available

Data Directories

The Internet's offerings for data, software, and other primary resources are vast and quite diverse. Comprehensive data directories, such as the Global Change Master Directory (GCMD)[5], make it easier for researchers to find and locate data relevant to specific research needs. As with other such data directories on the Internet, GCMD offers descriptions of data-gathering projects, inventories of data holdings, and descriptions of individual datasets.

Central Databanks

Large, unified databanks provide central storage of interrelated, complex data, adding value to the holdings by adding uniform structure, indexing, and multiple options for search and retrieval. Probably the best known of the large databank services is GenBank, the Genetic Sequence Data Bank maintained and supported by the National Center for Biotechnology Information (NCBI) to store, manage, and distribute genetics data and other information in support of the Human Genome Project. The large size and complex nature of sequence data generated by researchers in this field precludes "sharing" on the printed page, encouraging workers to deposit their data in central databases such as GenBank as part of the publication process. As a result, biologists now routinely reference central databanks such as GenBank in their published articles.

Individual Data Sets

The bridge between publishing and data sharing is also important for improving access to small, individual data sets. The smaller, specialized nature of these data sources makes them difficult to discover and locate. But if they are considered as supplementary material to a published article, their location may be referenced in print, and they may be loaded onto the public archive site of the journal's publisher. For example, backup data for articles published in the *Journal of Fluids Engineering* are routinely archived on Internet servers at the Scholarly Communications Project of Virginia Tech.[6] The American Chemical Society also archives supplementary material from selected journals on their Gopher site.[7]

Software Repositories

Software, source code and other executable files, as well as their accompanying documentation, is also available from numerous software archives on the Internet. Some archives are developed specifically for one type of computer platform, e.g., the Stanford Sumex archive for

SCIENCE AND TECHNOLOGY ON THE INTERNET

Macintosh software, the Naval Research Lab's Macintosh Science and Technology Archive, and SimTel archives for MS-DOS software.[8] Others repositories, such as the Netlib site maintained by the University of Knoxville and Oak Ridge National Laboratories,[9] provide software and other resources used by a particular discipline without regard to platform.

Multimedia Libraries

The last type of primary source available for scientific or technical application on the Internet are the libraries of multimedia files, used in research areas that rely on high-quality multimedia: photographs, sound clips, visualizations, 3D animations, and more. Examples of these collections include the AVES archive of bird images and sounds,[10] and the WWW-accessible astronomical archive of images, spectral data and more.[11]

 23

Methods for Sharing and Accessing Data and Software Resources

Many of the tools currently used on the Internet were specifically developed to facilitate data and resource sharing between scientists, and are still effective for that purpose. For further information about most of the tools described below, refer to the corresponding Fact Sheets in Section B of this guide.

Electronic mail

Though traditional email supports only text-based communication, programs that support the new standard MIME (Multiple Internet Mail Extensions) allow users to attach and send a binary file through electronic mail. This method provides a quick and convenient means of exchanging data between two colleagues.

Mailservers

Mailservers—that is, servers based on the email system, enable researchers to submit search requests in the form of an email message. Some mailservers programs, such as the blast and retrieve services available for searching GenBank and other sequence databases, support sophisticated queries that include boolean logic, search limits by field, and more. The drawback to data searches by mailserver are the lack of interactivity and the slow response time, meaning that a researcher may need to wait a short while to receive search results.

Telnet

Interactive databases available on the Internet are often accessed with Telnet either directly, or through a gateway from a Gopher menu or World-Wide Web page. The difficulty in using these databases is not in making the Telnet connection, but in learning the particular commands and search options available for that particular system. It is therefore important to follow the instructions on the screen and use online help (typically invoked with 'help' or '?'), where available.

File Transfer Protocol

File Transfer Protocol, or FTP, is one of the original Internet protocols developed to facilitate data exchange and resource sharing among scientists. It is still commonly used for this purpose.

You may use anonymous FTP to retrieve datasets, software, user documentation, and various other files from public-domain sites. You may also use FTP to move files between the computers of known colleagues (usernames and passwords are required).

Using FTP, you will encounter a wide variety of file formats, reflecting the diverse types of data files, software and other executables, images, animations and sound clips available to sci/tech users of the Internet. You may need to download and process FTP files to make them useful on your own desktop computer.

For information on handling the most common scientific data formats, refer to Ilana Stern's "Sci.Data.Formats FAQ" referenced below. See also the File Formats Fact Sheet in Section B of this guide.

WAIS

WAIS is a search tool that is useful for data collections that have been indexed, whether the data is in text, image, or another format. Using WAIS client software, one may search multiple servers at once. WAIS supports most any type of search string, including text, genetic sequence, latitude and longitude, and more.

WAIS users may encounter different versions of the software (e.g., simple WAIS, freeWAIS, and IUBio WAIS), with differing capabilities. Of the versions most commonly found, IUBio WAIS offers the greatest level of sophistication, supporting the logical operators AND, OR, and NOT, phrase searching, and truncation.

You may search WAIS directly, or through a gateway from a Gopher menu or World-Wide Web page.

Gopher

Both Gopher and the World-Wide Web are quick and easy interfaces offering users the ability to browse and search through menus of available resources before selecting and download specific files of interest. The WWW offers the additional capability of 'using' resources online: viewing photographs, running animations or visualizations, listening to sound clips, and so on. As users grow accustomed to 'one-stop shopping' on Gopher or WWW servers, these access methods are quickly gaining application as central starting points for searchers.

Specialized Software Agents

Finally, the many specialized Internet access tools developed for convenient search and retrieval fall into the general category of "software agents". These tools facilitate ready access to one or more databank services through a convenient 'front end' installed on the user's computer. Using network-based, client-server architecture, data agents take the search request of the user, communicate with search servers at the various databank sites, and then store and return the results to the user's desktop for immediate review. The program Entrez, developed by the National Center for Biotechnology Information[12] for searching gene sequence databases, is an example of this type of tool.

 24

Quality and Value

Data, software, and related resources are generally made available to the public 'as-is'. It is chiefly in the hands of the investigator(s) who generated these resources to ensure their validity and accuracy. In larger data collection projects, such as the genetics databanks, involvement of a moderator or curator may serve to screen contributions for relevance and possibly for consistency and conformance to relevant standard as well. At the mathematical software repository Netlib, software submissions are refereed, but the maintainers do not guarantee the quality of the software available on their site. As stated in the informational file for Netlib "...the lack of bureaucratic, legal, and financial impediments encourages researchers to submit their codes, knowing that their work will be made freely available quickly to a wide audience for testing as well."[13]

With software and other executable files available from public archives, there is also concern for possible virus infection. The maintainer at each site often includes a disclaimer concerning the possible risks to users of public-domain software.

How to Find Data and Software Resources in Your Area

The best guide covering available documentation and software for the most scientific data formats is Ilana Stern's "Scientific Data Format Information FAQ," regularly posted to the Usenet group sci.data.formats and archived at the MIT Usenet FAQ archive site:

```
ftp://rtfm.mit.edu/pub/usenet/news.answers/
sci-data-formats
```

The most popular data and software sites for any given field are also typically included on the subject hubs for that subject area. For example, the Computation Biology gopher at Johns Hopkins University

```
gopher://gopher.gdb.org
```

contains links to over 70 FTP sites with data and software for biology. The WWW biology subject hub at Harvard University offers a link to the 'Biologists Search Palette', providing a single point of access to "the most useful search engines for biological databases on the Internet"

```
http://golgi.harvard.edu/contrib/palette.html
```

For a list of subject hubs in the various disciplines of science and technology, or for more guidance in finding resources of any kind to meet your needs, refer to Section C of this guide.

Endnotes

1. Green, David G., "Database Diversity—A Distributed Public-Domain Approach." *Taxon,* vol. 43, no. 1 (February 1994): 51-62.

2. These collaboration tools are discussed in the module "Conferencing and Collaboration," in Section A of this Guide.

3. For brief information about UCSD's "dial-a-microscope," see pages 60-61 in: Commission on Physical Sciences, Mathematics, & Applications Staff, *National Research Council, National Collaboratories: Applying Information Technology for Scientific Research.* Washington, D.C.: National Academy Press, 1993.

4. Midell, David A., "Images from the Deep." *Byte* (June 1990): 256-260.

5. The Global Change Master Directory, produced by the cooperative efforts of NASA, NOAA, USGS, NCAR, and other data-gathering agencies is available from:

    ```
    telnet://gcmd.gsfc.nasa.gov (username gcdir)
    ```

6. *Journal of Fluid Engineering* databank is available from the Scholarly Communications Project of Virginia Tech:

    ```
    ftp://borg.lib.vt.edu/pub/JFE
    ```
    or
    ```
    gopher://borg.lib.vt.edu:70/11/jfe
    ```

7. The American Chemical Society's archive of supplementary materials is available from the ACS Gopher:

    ```
    gopher://acsinfo.acs.org:70/11/Publications/supmat
    ```

8. The locations of these three software repositories are, respectively:

```
ftp://sumex-aim.stanford.edu/info/mac
ftp://ra.nrl.navy.mil/MacSciTech
ftp://archive.orst.edu/pub/mirrors/simtel/msdos/
```

9. For more information about Netlib, send an email message to

```
netlib@ornl.gov
```

with the message:

```
send index
```

10. The AVES archive offers bird images and sounds compiled from the Smithsonian and other sources. It is available at

```
gopher://vitruvius.cecer.army.mil
```

11. For more information about astronomical archives accessible on the WWW, see the following article: Pasian, Fabio and Riccardo Smareglia, "WWW Access to Astronomical Archives and Databases." *International Journal of Modern Physics*, vol. 5, no. 5 (1994): 817-830.

12. A clear description of Entrez is available on pages 49-51 in: Commission on Physical Sciences, Mathematics, & Applications Staff, *National Research Council, National Collaboratories: Applying Information Technology for Scientific Research.* Washington, D.C.: National Academy Press, 1993.

13. This quote was excerpted from the file "General Information About NETLIB" available from:

```
gopher://gopher.ornl.gov:70/00/TechResources/
lib_net/GenInfo
```

Sharing Data Exercise 1. LOCATE, RETRIEVE AND PROCESS A SOFTWARE
APPLICATION FOR MACINTOSH

Task You are an engineer wanting to locate and retrieve a copy of MacWave, a
digital timing diagram editor needed to perform simple digital simulations.

Approach First you need to use Archie to locate any public archives sites that offer
copies of MacWave. You may then use FTP to retrieve the file to your
account, and download it to your Macintosh. More than likely, you will also
need to process the file before running the application.

Connect to one of the public
Archie clients, such as the
one at Rutgers.

```
% telnet  archie.rutgers.edu
Connected to archie.rutgers.edu.
Escape character is '^]'.

login: archie
# Bunyip Information Systems, 1993, 1994
# Terminal type set to `vt100 24 80'.
# `erase' character is `^?'.
# `search' (type string) has the value `sub'.
```

Enter your email address
for delivery of results.

A search type of 'sub'
means Archie will look for
partial or exact matches to
the search string, without
regard to case. Since you
are uncertain of the exact
name and version of the
application, you should
stick with this type of
search.

```
archie> set  mailto  clementg@solix.fiu.edu
archie> prog macwave
# Search type: sub.
# Your queue position: 12
# Estimated time for completion: 1 minute, 50 seconds.
working... =

Host faui43.informatik.uni-erlangen.de    (131.188.1.43)
Last updated 00:50 30 Dec 1994

    Location: /mounts/epix/public/pub/Mac/misc/compsci
```

When results are ready,
they scroll past the screen
too quickly to read.

```
.
.        <display deleted>
.
```

Request that results be sent
via email.

```
archie> mail
archie> quit
# Bye.
```

Back on your own computer
account, you receive an
email message from
Archie.

```
archie [prog macwave] part 1 of 1
.
.
.
```

The message contains the
results of the Archie
search.

```
Host ra.nrl.navy.mil    (132.250.1.121)
Last updated 06:52  7 Jan 1995

Location: /MacSciTech/elecEng
```

After considering the age of the available files, and the reputation of the organizations providing them, you choose the Naval Research Lab's Macintosh Science and Technology Archive.

From the information provided by Archie, you know exactly what site, directory, and file you need.

You may leave the transfer mode on the default, ASCII, because the desired file has been 'binhexed' (translated from binary to hexadecimal).

Check your home directory to see that the file is there.

Use the desired protocol (in this case, zmodem) to transfer the file from your shell account to your desktop computer.

The remaining steps depend on whatever decompression/translation package is available on your desktop computer. This exercise demonstrates the use of Aladdin Systems's freeware package, Stuffit Expander.

```
FILE -rw-r--r--123811 bytes 19:00 31 Jan 1993 MacWave1.2.sit.hqx

% ftp  ra.nrl.navy.mil
Connected to ra.nrl.navy.mil.
Name (ra.nrl.navy.mil:clementg): anonymous
331 Guest login ok, send ident as password.
Password: clementg@solix.fiu.edu
230 Guest login ok, access restrictions apply.
ftp> cd  /MacSciTech/elecEng
250 CWD command successful.
ftp> dir
200 PORT command successful.
150 ASCII data connection for /bin/ls (131.94.128.200,3232)...
-rw-r--r--  1 22 78   613 Jun  2 1992 CircuitDrawings.ReadMe
-rw-r--r--  1 22 78 43451 Jun  2 1992 CircuitDrawings.sea.hqx
-rw-r--r--  1 22 78 123811 Feb  1 1993 MacWave1.2.sit.hqx
.
.      <display deleted>
.
-rw-r--r--  1 22 78 255240 Sep 17  1991 pcb-cad.forPC.hqx

226 ASCII Transfer complete.
1663 bytes received in 0.31 seconds (5.3 Kbytes/s)
ftp> get  MacWave1.2.sit.hqx
200 PORT command successful.
150 ASCII data connection for MacWave1.2.sit.hqx
226 ASCII Transfer complete.
local: MacWave1.2.sit.hqx remote: MacWave1.2.sit.hqx
123812 bytes received in 1.6 seconds (75 Kbytes/s)
ftp> quit
221 Goodbye.

% dir
total 3392
-rw-------  1 clementg   123812 Jan 30 10:19 MacWave1.2.sit.hqx
drwxr-xr-x  2 clementg      512 Apr  9  1994 Mail/
drwxr-xr-x  2 clementg      512 Dec 18 09:15 News/

% sz  MacWave1.2.sit.hqx
sz: 1 file requested:
MacWave1.2.sit.hqx
Sending in Batch Mode

### Receive (Z) MacWave1.2.sit.hqx: 123812 bytes, 1:15 elapsed,
1646 cps, 85%
------------------------
```

Drag the file **MacWave1.2.sit.hqx** onto the icon for **Stuffit Expander**.

The file is automatically translated back to binary to become **MacWave1.2.sit**.

Drag **MacWave1.2.sit** onto the **Stuffit Expander** icon to produce the useable application file, **MacWave1.2**.

Sharing Data Exercise 2. LOCATE, RETRIEVE, AND PROCESS SOFTWARE
FOR THE PC.

Task You are a marine scientist and want to retrieve a copy of the tidal prediction
program, TIDES, for use on your PC.

Approach Since you know the exact version and name of the file (TIDES305.ZIP),
you may run an exact-match search in Archie and quickly find the location of
the program. Then you may retrieve the file using FTP, and download it
from your shell account to your PC. There you must unzip the file with one
of the many file conversion/processing packages available for PC s.

Connect to one of the public
Archie clients, such as the
one at InterNIC

(For alternate access
instructions using WWW,
go to the end of this
Exercise)

```
% telnet  archie.internic.net

Connected to ds.internic.net.
Escape character is '^]'.

            InterNIC Directory and Database Services

Welcome to InterNIC Directory and Database Services provided by
AT&T.These services are partially supported through a
cooperative agreement with the National Science Foundation.

First time users may login as guest with no password to receive
help.

Your comments and suggestions for improvement are welcome, and
can be mailed to admin@ds.internic.net.

AT&T MAKES NO WARRANTY OR GUARANTEE, OR PROMISE, EXPRESS OR
IMPLIED, CONCERNING THE  CONTENT OR  ACCURACY OF THE  DIRECTORY
ENTRIES AND DATABASE  FILES  STORED  AND  MAINTAINED  BY  AT&T.
AT&T EXPRESSLY DISCLAIMS AND EXCLUDES ALL EXPRESS WARANTIES AND
IMPLIED WARRANTIES OF MERCHANTABILITY AND FITNESS FOR A
PARTICULAR PURPOSE.
```

Log in as archie

```
login: archie

* * * * * * * * * * * * * * * * * * * * * * * * * * * * * * * * * * * * * * * * * * * * * * * * * * * * * *

        Welcome to the InterNIC Directory and Database Server.

* * * * * * * * * * * * * * * * * * * * * * * * * * * * * * * * * * * * * * * * * * * * * * * * * * * * * *

# Bunyip Information Systems, 1993, 1994

# Terminal type set to `vt100 24 80'.
# `erase' character is `^?'.
# `search' (type string) has the value `sub'.
```

Because you know exactly
what you are looking for,
you may change the search
type from the default of
' sub' to 'exact'.

Archie then searches only
for exact matches *including*

case to the query.
Also, to avoid getting a long listing of all locations where this version of TIDES resides (after all, just one will do), set the maximum number of hits.

Archie quickly returns the location and path for the desired program.

Now you are ready to retrieve the file from the public archive site at University of Virginia.

Notice that this site enables you to convert/process some files before retrieving (this applies for files processed with the UNIX tools GNU Zip, Compress, and TAR)

However, the TIDES program may not be processed in this manner. Since it been compressed with some version of the DOS application ZIP, it must be unprocessed with the corresponding application UNZIP on the PC.

Switch to binary mode before initiating the transfer.

Once the file has been transferred to your shell account, you are ready to download it to your PC.

```
archie> set search exact
archie> set maxhits 5
archie> prog TIDES305.ZIP
# Search type: exact
# Your queue position: 1
# Estimated time for completion: 2 minutes, 20 seconds.
working... =

Host ftp.virginia.edu    (128.143.2.7)
Last updated 00:26 16 Jan 1995

Location: /public_access
FILE   -rw-rw-rw-  116731 bytes  21:26 20 Oct 1994  TIDES305.ZIP

(END)
archie> quit
# Bye.
Connection closed by foreign host.

%ftp  ftp.virginia.edu
Connected to ftp.virginia.edu.
220 uvaarpa.Virginia.EDU FTP server (Version wu-2.4(9)...
Name (ftp.virginia.edu:clementg): anonymous
Guest login ok, send your complete e-mail address as password.
Password:clementg@solix.fiu.edu
Welcome to ftp.Virginia.EDU.  Please report trouble to
wfp5p@virginia.EDU.
.
.    <display deleted>
.
This ftp server supports compress/uncompress on the fly.  If
you do "get" <directory>.<suffix> or <file>.<suffix> where
suffix comes from the set [gz, Z, tar, tar.Z, tar.gz] the
server will compress and/or tar it for you on the fly. Likewise
if you strip suffixes from filenames, they will be uncompressed
for you.

ftp>cd  public_access
uvaarpa.virginia.edu:/public_access
ftp> binary
Type set to I.
ftp>get  TIDES305.ZIP
Receiving file: TIDES305.ZIP
100%  0 116731 bytes. ETA:  0:00

TIDES305.ZIP: 116731 bytes received in
ftp> quit


%  sz  TIDES305.ZIP
**
### Receive (Z) TIDES305.ZIP: 116731 bytes, 1:08 elapsed, 1697
cps, 44%
```

The final step depends on the particular file processing application available on your PC. This exercise demonstrates the use of UNZIP.EXE (available as the self-extracting archive file 'UNZ512X.EXE ' from the MS-DOS software archive at:

```
ftp://archive.orst.
edu/pub/mirrors/
simtel/msdos/zip/
```

ALTERNATIVE ACCESS METHOD FOR ARCHIE USING W W W

Connect to the WWW Archie Gateway at Rutgers.

Accessing Archie from WWW enables you to select and retrieve desired files on the spot, instead of having to initiate a separate FTP session after exiting Archie.

Also note that the WWW page provides a convenient menu of selections for search options, so that you don't have to key in commands for each setting.

Back on the PC, type :

```
A> unzip    TIDES305.ZIP
```

```
%  lynx  http://www-ns.rutgers.edu/htbin/archie
```

Rutgers University Archie Service (p1 of 3)

RUTGERS UNIVERSITY ARCHIE SERVICE

_____

Archie locates publicly available software and other information on the Internet. This WWW gateway will let you search for and retrieve files for use on your local computer.

Enter the full or partial name of the software or file for which you are searching in the Search String field below. You may set other parameters for the search, though the default settings are recommended. To start the search, press Enter or use the Start Search field.

Please note that the Archie server may take several minutes to complete its search.

The results from the search will be a list of files that match the search criteria. Click on the desired files to download them to your local computer.

_____

Archie Server: [archie.rutgers.edu_]

Sort responses by:
     (*) Date
     ( ) Host

Search for:
     ( ) Matching Substring (Ignore Case)
     ( ) Matching Substring
     ( ) Regular Expression Match
     (*) Exact Match

Sharing Data Exercise 3. USE NETLIB TO FIND MATHEMATICAL SOFTWARE

Task You are interested in finding some routines for data manipulation, but are only interested in "high quality" software.

Approach Check the NETLIB software repository maintained by the University of Tennessee and Oak Ridge National Laboratory. According to its information files, NETLIB offers a large collection of high-quality public-domain mathematical software.

You may obtain help in using NETLIB by sending email to an alias email account at Oak Ridge.

(For alternate access to NETLIB information files using Gopher, see the sample session at the end of this exercise).

```
% pine

   PINE 3.90    MAIN MENU     Folder: INBOX   0 Messages


   ?  HELP               -  Get help using Pine
   C  COMPOSE MESSAGE    -  Compose and send a message
   I  FOLDER INDEX       -  View messages in current folder
   L  FOLDER LIST        -  Select a folder to view
   A  ADDRESS BOOK       -  Update address book
   S  SETUP              -  Configure or update Pine
   Q  QUIT               -  Exit the Pine program



Copyright 1989-1994.  PINE is a trademark of the University of
Washington.

C

PINE 3.90    COMPOSE MESSAGE  Folder: INBOX   0 Messages

To      : netlib@ornl.gov
Cc      :
Attchmnt:
Subject :
----- Message Text -----
send  index

^G Get Help  ^X Send       ^R Read File ^Y Prev Pg   ...
^X


To: clementg@solix.fiu.edu
Subject: Re: send index   1
.
.      <display deleted>
.
 ===== How to use netlib =====
 This file is the reply you'll get to:
        mail netlib@ornl.gov
        send index
```

Compose and send a mail message using the commands available in your mail program.

You will soon receive a message back from netlib with instructions on searching the archive, descriptions of holdings, etc.

```
Here are examples of the various kinds of requests.
*  get the full index for a library
        send index from eispack
*  get a particular routine and all it depends on
        send dgeco from linpack
*  get just the one routine, not subsidiaries
        send only dgeco from linpack
*  get dependency tree, but excluding a subtree
        send dgeco but not dgefa from linpack
*  just tell how large a reply would be, don't actually send
    the file
        send list of dgeco from linpack
*  get a list of sizes and times of all files in a library
        send directory for benchmark
*  search for somebody in the SIAM membership list:
        who is gene golub
*  keyword search for netlib software
        find cubic spline

    <display deleted>

-------quick summary of contents---------
a - approximation algorithms
aicm - selected material from Advances in Computational
Mathematics
alliant - set of programs collected from Alliant users
amos - special functions by D. Amos.  = toms/644

    <display deleted>
```

ALTERNATIVE ACCESS METHOD FOR NETLIB USING GOPHER:

Connect to the Gopher at Oak Ridge National Laboratories, and select the menu for Netlib.

Note that you may retrieve information files about the repository from this menu, but may not actually connect to the archives from here.

```
% gopher  gopher.ornl.gov
--> Technical Resources/
     --> Information about NETLIB: Math Routines/


        Internet Gopher Information Client 2.0 p110

        Information about NETLIB: Math Routines

    1.  General Information About NETLIB.
    2.  How to Use NETLIB.
    3.  Quick Summary of Contents.
    4.  A Bit More Details.
    5.  NETLIB at UTK/ORNL/
```

Sharing Data Exercise 4a. DOWNLOAD AN IMAGE FROM THE WEB.

Task As an astronomy professor, you want to get an image of the collision between Comet Shoemaker-Levy and Jupiter to show students in a lecture.

Approach Many comet images are available on the World-Wide Web. Using a text browser such as Lynx, you will only be able to read a description of what's available, and then retrieve any desired images for display offline. With a graphical browser (as explained at the end of this exercise), you can view images online, enabling you to browse before saving.

Point the Lynx browser to the Comet images site at Goddard Space Flight Center.

Note that, in place of an image that would be displayed with graphical Web browsers, Lynx 'sees' a text description of the image instead.

```
% lynx   http://nssdc.gsfc.nasa.gov/planetary.comet.html

                      Comet Shoemaker-Levy 9 (NSSDC) (p1 of 4)
_____


    Image of Comet P/Shoemaker-Levy 9

           COMET P/SHOEMAKER-LEVY 9 COLLISION WITH JUPITER
_____

Do not go gentle into that good night.
Rage, rage against the dying of the light. - Dylan Thomas


 From July 16 through July 22, 1994, fragments of Comet
 P/Shoemaker-Levy 9 collided with Jupiter, with dramatic
effect. This was the first collision of two solar system bodies
ever to be observed. Shoemaker-Levy 9 consists of 20
discernable fragments with diameters estimated at up to 2
kilometers, which impacted the planet at 60 Km/s. The impacts
resulted in plumes many thousands of kilometers high, hot
"bubbles" of gas in the atmosphere, and large dark "scars" on
the atmosphere which have lifetimes at least on the order of
weeks. Smaller bits and dust continue to impact the planet.

Shoemaker-Levy 9 is gone, but as the Earth- and space-based
images show, it did not go quietly.

  * Images of the Collisions
  * The Comet
  * The Impact

    .
    .        <display deleted>
    .
```

Select the collection of collision images.

IMAGES FROM COMET P/SHOEMAKER-LEVY 9 COLLISION WITH JUPITER

   This is the way the world ends
   not with a bang, but a whimper. - T.S. Eliot

   This may be how worlds end, but certainly not comets, as the
following images demonstrate.

* Image index with brief summary

* Image index by fragment

* Galileo - Fragments K, N, W [Updated: Dec 5]
* **Hubble - Fragments A, C, E, G, K, P, Q, R** ...
* ESO - Fragments A, D, E, F, G, H, K, L, R, S ...
* Calar Alto - Fragments A, C, D, E, F, G, H, Q, ...

   .
   .   <display deleted>
   .

HUBBLE IMAGES FROM COMET P/SHOEMAKER-LEVY 9 COLLISION WITH
JUPITER

For any of the images, click on the image to view the original

The images are provided courtesy of the HST Science Team

   Fragment A - July 16

   [IMAGE] [IMAGE] [IMAGE]
   * Caption
   [IMAGE]
   * Caption
   [IMAGE]
   * Caption

When a specific image file is selected, Lynx is unable to display the image on the screen.

This file cannot be displayed on this terminal:
D)ownload, or C)ancel **D**

   Arrow keys: Up and Down to move. Right to follow a link; Left
to go back.

Instead, Lynx gives you the option of saving the file on your own computer.

 H)elp O)ptions P)rint G)o M)ain screen Q)uit /=search

DOWNLOAD OPTIONS

You may name the file as desired, and exit Lynx.

You have the following download choices please select one:

Save to disk

No other download methods have been defined yet. You may define an unlimited number of download methods using the lynx.cfg file.

Enter a filename: **sl9a_hst1.gif**

Arrow keys: Up and Down to move. Right to follow a link; Left to go back.

H)elp O)ptions P)rint G)o M)ain screen Q)uit /=search [delete]=history list **q**

Back on your computer account, you may download the file to your desktop computer using Zmodem or whatever file transfer protocol is supported by both the sending and receiving computer.

The GIF file is now available on your desktop computer. It may be viewed with a number of viewing packages, such as JPEG View, or converted to another graphics format that is compatible with many paint and publishing programs).

% sz   sl9a_hst1.gif

sz: 1 file requested:
sl9a_hst1.gif

Sending in Batch Mode
*
Receive (Z) sl9a_hst1.gif: 138878 bytes, 1:24 elapsed....

Sharing Data Exercise 4b. VIEW AN IMAGE FROM THE WORLD-WIDE WEB 'ONLINE'.

If you have direct Internet access and a graphical WWW browser such as NetScape, Mac- or WinWeb, Mosaic, or Arena, you may view images online before saving them to disk. The commands and capabilities for performing this function vary from one client to the next, but in general follow these steps:

1. Open the URL
`http://nssdc.gsfc.nasa.gov/planetary/comet.html`

2. Select the desired collection of images by clicking with the mouse.

3. Individual images that are available for viewing and downloading are denoted by either an icon, or a 'fingernail' or miniature version of the image.

4. To call up the full-sized version of the image, click on the icon or fingernail. A connection will be made to the appropriate image file, and the image will appear (sometimes it takes a few minutes to transfer, because image files tend to be rather large!).

5. With some clients, such as MacWeb, you save the image by holding down the mouse until a menu appears with an option for **'Retrieve to Disk'**. You may then enter a name for the image file, select **'Save'**, and the transfer will take place.

6. With some clients, such as Netscape, you transfer the image file to your desktop computer by invoking the **'Save '** option under the **'File'** menu, selecting the option for **'Save as Source'**, and naming the image file as desired.

7. Once the image file is has been transferred, you may view it or convert it to another format with a number of software packages including JPEG View, Graphic Workshop, Graphic Converter, Imagery, and many others.

Section B: Fact Sheets

Archie

Email

File Formats

Finger

FTP

Gopher

HTML

HYTELNET

IRC

Jughead

Mailservers

MBone

MIME

MOO

Network News

Telnet

URL

Veronica

WAIS

Web Robots

WWW

Archie

What Is Archie?

Archie is a search tool that helps users identify and locate all types of files accessible from anonymous FTP archives on the Internet. Based on the user's search terms, Archie looks for matches in the filenames and directory titles of hundreds of anonymous FTP sites around the Internet.

Archie is a client-server application developed by Peter Deutsch and colleagues at McGill University. The user enters the desired search terms in an Archie client with the 'prog' command, and waits for Archie to return a listing of FTP sites from which the file may be retrieved, and the directory, file name, date of last file revision, etc. Archie also includes a database containing brief descriptions of many of the files and programs available on FTP archives; this database is searchable with the 'whatis' command.

Users may run an Archie client residing on their own Internet host, or one of the public Archie clients accessible through Telnet. Additionally, a menu-driven interface to Archie may be found on select Gophers and World-Wide Web servers. Archie searches may also be initiated in "batch mode" by sending a search request to any Archie server via electronic mail. The latter technique is often the most efficient approach because Archie searches can, at times, take several minutes to complete, typing up one's computer until the processing is done.

Archie offers a number of search options that allow users to sort results, limit the number of hits, send results to an electronic mail box, etc. In interactive search mode, one may view all available options and their default settings with the 'show' command. Those searching Archie through electronic mail may request a list of settings by inserting the command 'help' in the mail message.

Applications That Use Archie

- Identify and get the exact location, filename and other information for any file accessible using anonymous FTP, including texts, software, graphics, and more.

- Obtain brief descriptions of software and other files available by anonymous FTP.

How to Start Archie (the Public Client)

```
%          telnet archie.sura.net
login:     archie
archie>    show
           # 'encode' (type string) has the value 'none'.
           #'language' (type string) has the value 'english'.
           #'mailto' (type string) is not set.
```

   <display deleted>

```
archie>    prog vicar
```

   <display deleted>

```
archie>    mail
archie>    quit
           #Bye
Connection closed.
```

How to Use Archie (By Mail)

```
%          mail archie@archie.sura.net
Subject:

           set mailto clementg@solix.fiu.edu
           set maxhits 10
prog       vicar
```

Exercises Demonstrating Applications of Archie
Publishing Exercise 3
Sharing Data Exercises 1 and 2

For Further Information:

Robison, David F.W., "Archie", *All About Internet FTP* (p. 21-24)
Covers Archie commands and search options.

Electronic Mail

What Is Electronic Mail?

Electronic mail, or email, is the most popular tool on the Internet, used by millions every day to communicate with other email users around the world. You may send an email message directly to any other user on the Internet and, through an Internet gateway, to a user on another network as well.

What makes email so ubiquitous and powerful is its underlying protocol, the Simple Mail Transfer Protocol (SMTP). One of the original Internet protocols, SMTP describes how email messages should be formatted, transmitted, and delivered. Although there is a plethora of email programs in use on the Internet, SMTP ensures that they are all universally capable of sending and receiving messages to and from each other.

Email programs on the Internet share a common set of capabilities: send, receive, reply, forward, attachment of a text file, and mail management features such as folders. Other common email options may include a cc: line (for sending a "carbon copy"), a blind cc: option (for sending a "blind carbon copy" when you don't want the main recipient to know you're sending a copy to others), aliasing of addresses, and distribution lists for mass mailings. Sending binary files is also possible in some electronic mail programs (see the Fact Sheet for MIME in this section.)

Sending or receiving electronic mail is usually a simple operation, and unsuccessful email attempts are most often due to incorrect addressing. In general, an Internet address follows the syntax "user@host" (*e.g.*, "clementg@solix.fiu.edu"), where "user" is the individual's username on the Internet host and "host" is the name of the computer holding the account. When mail can not be delivered to another user, it is generally because either (or both) of these components is not accurate. For example, if someone incorrectly addresses a mail message to clementg@solax.fiu.edu the daemon on their system will soon return the message with the subject "Returned mail. Host unknown." A message addressed to clement@solix.fiu.edu will be returned with the subject "Returned mail. User unknown." Only those messages properly addressed to clementg@solix.fiu.edu arrive in the appropriate email box, and they usually do so within seconds or minutes of being sent. For strategies and techniques for finding someone's proper email address on the Internet, see the "Finding Colleagues" module in Section A of this guide.

Another possible delivery problem may result from email going to a network outside the Internet. For mail to BITNET, for example, messages must be routed through an "Internet-BITNET gateway," that is, a computer that sends mail on both networks. Addressing for such messages may take two possible forms (one or both may be supported on any given Internet node):

```
        username%computer.bitnet@gateway
e.g.    clementg%servax.bitnet@cunyvm.cuny.edu

        gateway!computer.bitnet!username
e.g.    cunyvm.cuny.edu!servax.bitnet!clementg
```

In the above examples, **cunyvm.cuny.edu** is one available Internet-BITNET gateway; others include

```
        cornellc.cit.cornell.edu
        vm1.nodak.edu
        pucc.princeton.edu.
```

One should try and use the closest possible gateway.

On computers connected to BITNET, or another network besides the Internet, the mail system may require users to prefix an email address with the protocol to be used for delivery. A message going over the Internet might be addressed as

```
        SMTP%"clementg@solix.fiu.edu
```

A BITNET address might be addressed as

```
        BITNET%"clementg@servax.bitnet
```

Once accustomed to the features of your own mail program, you will find a multitude of applications for electronic mail beyond personal correspondence.

Applications that Use Email

- Send correspondence, manuscripts, or other text files to another colleague

- Join, participate in, and search archives of Mailserver Discussion Groups

- Send search requests to other Internet servers (*e.g.* WAIS, Archie)

- Send search requests to a research database

- Subscribe to electronic newsletters, bulletins, or to Table of Contents services

Exercises Demonstrating Applications of Electronic Mail

Conferencing Exercises 1 and 2   Reference Exercise 2
Finding Colleagues Exercises 3 and 4   Sharing Data Exercise 3

For Further Information

Email programs vary greatly in their commands, so it is best to check the documentation for your particular mailer for further information.

File Formats

What File Formats Are Available on the Internet?

Sci/tech users of the Internet are likely to encounter a wide variety of file formats depending on the type of resource they are retrieving: simple or formatted text, image, motion video, sound, data, executable files, and more. While downloading these files is fairly simple using FTP, Gopher, or a WWW client, users may find it confusing and frustrating to learn how to process these files and make them usable on one's own computer.

To determine what technique or tool to use for processing the many different types of files on the Internet, one must first recognize the format of a network file by examining its extension. It is important to recognize that formatting can occur at three different points, with differing, and sometimes cumulative, results.

1. File creation

When a file is created and saved in any application, it is stored in a particular format. Text files created in word processing packages may be saved as simple text (ASCII) or as formatted text with embedded codes that control style, layout, or placement of graphics. Extensions may include .txt for simple text, .wp for WordPerfect files, .rtf for Rich Text Format files, and so on.

Sci/tech users of the Internet may also need to become familiar with extensions denoting formats used specifically for technical publishing. One such example is the TeX typesetting system, created (by mathematician Donald Knuth) to present mathematical formulae and other forms of scientific expression. TeX typesetting files are denoted by the extensions '.dvi' or '.tex'.

Files created and saved with other types of applications use other types of extensions: .xls for an Excel spreadsheet, .dbf for a Dbase file, etc. Of particular interest to scientists are the many formats used for scientific data, such as FITS (Flexible Image Transport System) or PDS (Planetary Data System). An extensive description of the various scientific data formats is available in the document "Sci.Data.Formats FAQ," listed in the 'For Further Information' section below.

2. Archiving and Compression

A group of related files may be archived into a single file so that they are stored and transmitted together on the network. For example, a text manuscript may be archived along with its tables and figures into a single file that represents a scientific article. The text file 'manuscript.tex' and its accompanying material 'table.ps' and 'image.gif' may be archived to become the file 'article.tar'.

Individual or archived files may also be compressed to take up less storage room and require less transfer time on the Internet. Compression may either replace or add another extension on to the original or archived file name. In the example of our scientific article, the compressed file would become 'article.tar.Z'.

Different compression and archiving programs are used for different computing platforms. PC files are often zipped (.zip) with pkzip; Macintosh files may be stuffed (.sit) with Stuffit; and Unix files may be archived (.tar) and then compressed (.Z) with the Unix compress command.

3. Encoding or Transformation

After creating a file in a given format, and then optionally compressing or archiving it, one may choose to encode or transform it for safe and accurate transmission over the Internet. A number of schemes are in use for the different computing platforms: uuencode (.uu) for UNIX; BinHex (.hqx) for Macintosh; Binary-to-ASCII (.btoa) on the PC; and others.

This step may either add an additional extension on to a file, or replace it, depending on the program being used. Binhexing a Macintosh file 'file.sit.' would produce 'file.sit.hqx'. Encoding our earlier scientific article ('article.tar.Z') with the Unix program 'uuencode' creates the file 'article.uu.'

Processing and Using Files

Files of any format may be transferred from the Internet to one's own computer and used there, as long as a suitable application is available to open and run them.

Files that have been encoded, compressed and/or archived must be processed first, to 'peel off' the layers of each format. For example, the file 'article.uu' would be processed as follows:

```
decode 'article.uu'          => creates 'article.tar.Z'
uncompress 'article.tar.Z'   => creates 'article.tar'
de-archive 'article.tar'     => creates 'manuscript.tex'; 'table.ps';
                                'image.gif'
```

For files that are usable on different computer operating systems, it shouldn't matter if the file retrieved from the Internet is processed with software from one computer platform, and then de-processed with software for another platform. For example, a Macintosh user retrieving a text file compressed with PKZIP on the PC (e.g., 'text.zip') may uncompress it with a Macintosh uncompression tool, such as ZipIt, and then import the file into a word processor on the Mac. There are also now available several utility programs capable of handling all types of compression and archive formats. For example, Stuffit Deluxe for the Macintosh can process and unprocess formats from UNIX and DOS platforms.

For an excellent summary of many file formats on the Internet and their requirements for transfer and processing, refer to Allison Zhang's "Multimedia File Formats on the Internet" listed in the 'For Further Information' section below. Also helpful is the document "File Compression, Archiving, and Text<->Binary Formats," a table that lists nearly every compression, archiving and translation scheme used on the Internet, the resulting file extensions, and the software needed to process the files.

Exercises Demonstrating File Processing

Publishing Exercise 2
Sharing Data Exercises 1 and 2

For Further Information

Cox, Jennifer and Mohamed Taleb, "Images On The Internet" *DATA-BASE* August 1994, 18-26.
Discusses popular image formats and software for viewing them.

Gaffin, Adam. "Odd letters — decoding file endings" Section 7.4 *EFF's Guide to the Internet* (formerly *The Big Dummy's Guide to the Internet*) January 6, 1995 Electronic Frontier Foundation

```
gopher://gopher.eff.org:70/11/Publications
Net_guidebooks/EFF_Net_Guide
```

Lemson, David. "File Compression, Archiving, and Text/Binary Formats"

```
ftp:// ftp.cso.uiuc.edu/doc/pcnet/compression
```

This extremely useful chart lists most every compression/archiving system in use on the Internet, the program(s) used to uncompress/unarchive for each operating system (DOS, MAC, Unix, Amiga, etc.), and the sites where such programs may be retrieved.

Macintosh Science and Technology Archive. "Accessing Files "

```
ftp://ra.nrl.navy.mil/MacSciTech/help/
accessing-files.txt
```

This brief guide provides basic information about retrieving, down-loading, and converting files to one's own Macintosh computer.

Petersen, Keith "README.file-formats"

> `ftp://archive.orst.edu/pub/mirrors/simtel/msdos/`
> `README.file-formats`

A concise guide to the possible extensions found on archived files for PC's, and the DOS-compatible packages used to process and unprocess them.

Robison, David F. W. *All About Internet FTP: Learning and Teaching to Transfer Files on the Internet.* Berkeley: Library Solutions Press, 1994.

This entire guide is devoted to FTP and covers details such as file types, compression and archiving techniques and software, etc. See the 'Common File Extensions' table on page 80.

Stern, Ilana. "Sci.data.formats FAQ Document" (continually updated)

> `ftp://rtfm.mit.edu/pub/usenet/news.answers/sci-data-`
> `formats`
>
> `http://fits.cv.nrao.edu/traffic/scidataformats/`
> `faq.html`
>
> `http://info.mcc.ac.uk:80/CGU/Visualisation/`
> `sdf.html`

This helpful guide provides documentation and pointers to software for the predominant scientific data formats available on the Internet.

Zhang, Allison. "Multimedia File Formats on the Internet" (continually updated)

> `http://ac.dal.ca/dong/contents.htm`

This is a comprehensive and clearly-written guide covering many types of file formats encountered on the Internet, what mode to transfer them in, and what tools to use for processing.

Some Common File Formats Used on the Internet

Text

.asc	ASCII.
.rtf	Rich Text Format
.tex	TeX Source File
.ps	PostScript®
.txt	Simple text
.src	WAIS Source File
.pdf	Portable Document Format (Adobe)

Still Images

bmp	Bitmapped (Microsoft Windows®)
.gif	Graphic Information Format (CompuServe)
.tiff	Tagged Image File Format
.jpeg	Joint Photographic Experts Group Format
.jfif	Joint Photographic Experts Group Format
.ps	PostScript
.eps	Encapsulated PostScript

Motion Video

.cgm	Computer Graphics Metafile
.dvi	Digital Video Interactive
.fli	Autodesk Animator®
.mpeg	Motion Picture

Sound

.aiff	Audio IFF
.au	
.mod, .nst	Amiga MOD Format
.snd	Various
.ul	ULaw
.wav	Waveform (Microsoft)
.voc	

Archiving, Compression, Encoding, Transforming Formats

.arj	ARJ		.shar	Shell Archive
.boo	BOO		.sit	Stuffit
.cpt	Compact Pro		.tar	UNIX Tape Archive
.dd	DiskDoubler		.uue	uuencode
.gz	GnuZip		.XXe	xxencode
.hqx	BinHex		.Z	UNIX Compress
.pit	Packit		.zip	PKZIP
.sea	Self-extracting File		.ZOO	ZOO

Finger

What Is Finger?

Finger is a simple network service primarily supported by computers using the Unix operating system, but sometimes available on other systems as well. Finger reports if a particular user is logged on (and if not, the date/time of last log on), and also displays any information included in the user's 'Plan'. A number of information providers use the 'Plan' to disseminate timely or summarized scientific information because the file is so easy to update or revise.

Applications That Use Finger

- Verify email address or activity of an individual's electronic mail account

- Obtain recent news or reference information included in a Plan

- Obtain current technical data included in a Plan

How to Use Finger

```
%finger quake@geophys.washington.edu
```

[geophys.washington.edu]

Login name: **quake**   In real life: Earthquake Information

Directory: **/u0/quake**   Shell: /u0/quake/run_quake

Last login   Tue Nov 8 17:13 on ttyi2

New mail received: Tue Nov 8 16:42:15 1994

 unread since: Tue Nov 8 00:48:39 1994

Plan:

```
The following catalog is is for earthquakes (M>2)
in Washington and Oregon produced by the Pacific
Northwest Seismograph Network, a member of the
Council of the National Seismic System.
```

   <display deleted>

DATE-TIME is in Universal Time (UTC) which is PST + 8 hours. Magnitudes are reported as local magnitude (Ml).
QUAL is location quality
     A-good
     D-poor
     Z-from automatic system and may be in error.

```
DATE- (UTC) -TIME  LAT(N)  LON(W)   DEP MAG QUAL        COMMENTS

yy/mm/dd  hh:mm:ss   deg.    deg.     km  Ml

94/10/06  21:30:42  45.66N  120.15W   1.3 2.7  C         66.1 km SSW of Prosser, Wa
94/10/11  19:45:14  44.10N  121.31W   3.8 2.2  B          5.6 km NNW of Bend, OR
94/10/12  15:56:36  44.61N  122.80W  22.3 2.1  C         40.5 km SSE of Salem, OR
94/10/13  11:13:23  46.35N  122.38W   7.9 2.5  B         12.8 km   W of Elk Lake
94/10/14  18:56:06  47.41N  122.81W   2.2 2.8  B    FELT 20.9 km   SW of Bremerton
94/10/17  22:50:40  48.06N  119.85W   2.3 2.0  C         28.2 km NNE of Chelan
94/10/18  16:00:19  48.53N  120.90W   1.0 2.3  D         51.7 km NNE of Glacier Peak
94/10/27  20:55:02  47.73N  119.91W   0.5 2.1  C         12.5 km SSE of Chelan
94/11/01  03:02:24  42.23N  121.95W   7.0 3.0  A    FELT 15.3 km   W of Klamath Fal
94/11/02  21:03:31  47.73N  119.93W   0.5 2.1  C         12.6 km SSE of Chelan
94/11/03  20:48:58  46.38N  119.25W   0.4 2.1  B    FELT 11.7 km NNE of Richland
```

Exercise Demonstrating Applications of Finger

Reference Exercise 1

FTP: File Transfer Protocol

What Is FTP?

File Transfer Protocol (FTP) is one of the original Internet protocols, developed to allow scientists on different computers to easily share files. Still widely used today, it enables users to quickly retrieve documents and files from publishers and other information providers, and to transfer files among collaborating colleagues.

There are two types of FTP in use on the Internet. Anonymous FTP allows any public user to login with the username 'anonymous' and use one's email address as the password. Users may move between directories, list available files, and retrieve one or more files with the **get** and **mget** commands, respectively. Many anonymous ftp sites have been set up on the Internet to provide publications and other files of general interest, shareware computer programs, etc.

The other type of FTP allows two users to transfer files they have agreed to share. It requires a known username and password.

Any type of computer on the Internet may be configured to allow ftp, but most sites use the UNIX operating system. Essential Unix commands for FTP users include:

cd	change directory	*get filename*	retrieve a file
pwd	print working directory	*get filename* **more**	print text file to screen
dir	full directory listing	**mget** *filename.*	*retrieve multiple files
ls	list files in a directory	*put filename*	add a file
ascii	use ascii transfer mode	**binary**	use binary transfer mode

A difficult part about using FTP is learning to recognize and work with the many types of file formats one finds available for FTP. Simple texts, encoded documents, photographs, video clips, audio files, executable software may all be archived on an FTP site and each requires a particular mode of transfer and possibly some additional steps to process for use on one's own computer.

At a minimum, FTP users need to know when to select ascii or binary transfer mode. Ascii is the default automatically used, and may be changed by typing **binary** before getting a file. Files requiring ascii transfer include simple texts (for example, .txt, .doc), some encoded texts (for example, Rich Text Format - .rtf; Postscript - .ps; Hypertext Markup Language-.html), and some files transformed from binary to ascii (.btoa, .hqx). Most other files require binary transfer.

For more information, see the "File Formats" Fact Sheet included in this section of this guide.

Applications That Use FTP

- Retrieve issues, articles, or supplementary material from electronic journals

- Retrieve software, data, sounds, maps, spreadsheets, or other files from public archives or colleagues

- Retrieve forms for proposal submission

- Retrieve formatted user manuals, tutorials, hypercard stacks

- Transfer one's own files and data to a colleagues' computer, and vice versa (password-controlled)

- Transfer files between own's own computers (work-to-home and vice versa)

Exercises Demonstrating Applications of FTP

Conferencing Exercise 3
News Exercise 5
Publishing Exercises 1 and 3
Sharing Data Exercises 1 and 2

For Further Information

Deutsch, P., Emtage, A., and Marine, A. "How to Use Anonymous FTP", Internet Engineering Task Force (IETF), Internet Anonymous FTP Archives (IAFA) Working Group, RFC 1635, FYI 24, May 1994.

```
gopher://nic.merit.edu:7043/00/internet/
documents/fyi/fyi_24.txt
```

Robison, David F. W., *All About Internet FTP: Learning and Teaching to Transfer Files on the Internet*. Berkeley, CA: Library Solutions Press, 1994.

Gopher

What Is Gopher?

Gopher is both an Internet protocol and a client/server program that enables users to access, browse, display, download, and search a variety of resources on the Internet. Through a series of hierarchical menus, users select resources that may reside on the same computer, or on another server elsewhere on the Internet.

Gopher interacts with a variety of different servers, including FTP, Archie, HYTELNET, WAIS, Netfind, Telnet, integrating information provided by these disparate servers into a single menu-driven interface. To identify the type(s) of information on a Gopher menu, users may need to refer to the following list of Gopher types :

0	—	Text File
1	—	Directory
2	—	CSO name server
3	—	Error
4	—	Mac HQX file.
5	—	PC binary
6	—	UUencoded file
7	—	Full Text Index (Gopher menu)
8	—	Telnet Session
9	—	Binary File
s	—	Sound
e	—	Event (not in 2.06)
I	—	Image (other than GIF) M — MIME multipart/mixed message
T	—	TN3270 Session
c	—	Calendar (not in 2.06)
g	—	GIF image
h	—	HTML, HyperText Markup Language

Gopher also seamlessly links to all other Gophers in the world, creating a rich information system known as 'Gopherspace.' Probably the hardest part about using Gopher is finding specific information in Gopherspace. To this end, two tools are available to perform keyword searches of Gopher menus—Veronica and Jughead. For more information on either tool, refer to the corresponding Fact Sheets in this section.

Some users may have a Gopher client running on their own Internet node; others may need to telnet to a publicly-accessible Gopher client available on several servers around the Internet.

Applications That Use Gopher

- Use various Internet clients (*e.g.*, Archie, Netfind, WAIS) through a menu-driven interface

- Search Phonebooks and other White Pages services to find people

- Search bibliographic, reference or numeric databases

- Browse and retrieve issues, articles, or supplementary material from electronic newsletters or journals

- Download software, data, sounds, images, maps or other binary files from archives

- Browse and retrieve user guides, tutorials, instructional materials

How to Use Gopher

To reach one of the (few remaining) public Gopher clients:

%	**telnet marvel.loc.gov**
login:	**gopher**

From a Gopher client on your own machine (no Gopher login needed):

%	**gopher mentor.lanl.gov**

Exercises Demonstrating Applications of Gopher

Conferencing Exercise 1
Finding Colleagues Exercises 1 and 6
News Exercises 1, 2 , 3 and 5
Reference Exercise 4
Literature Exercise 3
Publishing Exercises 1, 2, 4 and 6

For Further Information

"Gopher FAQ Document"

> **ftp://rtfm.mit.edu/pub/usenet/news.answers/
> gopher.faq**

"What is Gopher" archives at the Gopher 'mother site'
(University of Minnesota)

> **gopher://gopher.tc.umn.edu:70/11/
> Information%20About%20Gopher**

HTML: HyperText Markup Language

What Is HTML?

HTML is the standard formatting system underlying hypermedia documents on the World-Wide Web. HTML is one application of the broader Standard Generalized Markup Language (SGML), the international standard (ISO 8879) for marking up texts and other documents for electronic publication. If you want to create a home page or author a publication on the Web, you need to produce it in HTML format.

HTML documents are ASCII, or plain text, files, with embedded tags denoting basic style and format (*e.g.*, title, headings of various sizes, paragraph or line breaks, formatted lists, and so on). It is important to recognize, however, that HTML tags provide only minimal control over the ultimate appearance of a given document. Much of the formatting and style (*e.g.*, font, type size) is controlled by the user through his or her Web browser.

One of the primary features of HTML documents is the anchor or hyperlink—a 'hot' area of text that, when selected, calls up other files on the same computer or elsewhere on the Internet. These links may include other HTML documents, graphics, photos, sound, video, or information served by other Internet protocols such as Gopher, Telnet, and WAIS. The exact location, path, and necessary access method for each linked item is pinpointed by a Uniform Resource Locator (URL). For more information about URLs, see the corresponding Fact Sheet.

One may create a simple HTML document (text, images, and hyperlinks) with a standard text processor by inserting the tags manually. Alternatively, one may use one of the many HTML editors or converters that insert the requisite tags through a series of macros. Most of the software for creating or converting HTML files are detailed on the mother site at CERN.

More advanced applications of HTML, such as user input forms and clickable maps, require special scripting by an experienced Web developer.

The application of HTML for scientific publishing on the World-Wide Web has greatly elevated the functionality of electronic publications, enabling authors to connect related or explanatory information through hyperlinks, and to attach to the text related images, datasets, software, video, or sound. Several scientific journals on the World-Wide Web, including the *Journal of Artificial Intelligence Research*

```
http://www.cs.washington.edu/research/jair/
home.html
```

and the *Electronic Journal of Combinatorics*

```
http://ejc.math.gatech.edu:8080/Journal/
journalhome.html
```

demonstrate the possibilities of hypermedia science publishing on the Internet.

On the other hand, some scientific publishers find that HTML, in its current configuration, does not include sufficient elements and formatting options to effectively convey certain types of scientific information and data. For example, HTML currently does not integrate tables, mathematical formulas and other scientific notations into the text of the document. Rather, these items must be presented as images, displayed alongside the text.

To address these limitations of HTML, an enhanced version, known as HTML Level 3 or HTML+, is now under development. In fact, some WWW publishers are already incorporating features of HTML+ in their documents. Most users, however, are still limited to using hypertext documents authored with the original HTML specification because the common browsers (Mosaic, MacWeb, Netscape, etc.) support only that version. But a HTML+ compatible browser, Arena, is already in use (available only for XWindows systems) and is likely to become more widely available in the near future.

For the official specifications for HTML or HTML+, visit the WWW mother site at CERN

```
http://info.cern.ch/hypertext/WWW/MarkUp/
MarkUp.html
```

For information about Arena, consult CERN's Arena page

```
http://info.cern.ch/hypertext/WWW/Arena
```

Applications of HTML

- Online 'campus-wide' information systems for academic departments, universities and other research institutions, government agencies, publishers, businesses, and even individuals.

- Online publications such as reference books, newsletters, and scientific journals, combining text with multimedia 'figures' and hyperlinks to related information sources elsewhere on the Internet.

- Online museums combining narrative text with multimedia 'exhibits,' 'displays'

- Instructional modules combining explanatory text with visual and/or sound presentations and interactive exercises allowing user input

Exercises Demonstrating the use of HTML

Publishing Exercise 7

For Further Information About HTML

Andreessen, Marc. "A Beginner's Guide to HTML"

    `http://www.ncsa.uiuc.edu/demoweb/html-primer.html`

Flynn, Peter. "How to Write HTML"

    `http://kcgl1.eng.ohio-state.edu/www/doc/`
    `htmldoc.html`

Graham, Ian. HTML Documentation Site, University of Toronto

    `http://www.utirc.utoronto.ca/HTMLdocs/NewHTML/`
    `htmlindex.html`

Greene, Barr R. "WWW and HTML Developers' JumpStation"

    `http://oneworld.wa.com/htmldev/devpage/`
    `dev-page.html`

HTML/HTML+ collected resources and archives

    `http://info.cern.ch/hypertext/WWW/MarkUp/`
    `MarkUp.html`

Internet Engineering Task Force, "Hypertext Markup Language Specification 2.0", working draft.

    `gopher://nic.merit.edu:7043/00/internet/documents`
    `/internet-drafts/draft-ietf-html-spec-01.txt`

Powell, Jim. "Introduction to HTML"

    `http://scholar.lib.vt.edu/reports/soasis-slides/`
    `HTML-Intro.html`

Sullivan, Eamonn "Crash Course on HTML"

    `http://www.ziff.com/~eamonn/crash_course.html`

HYTELNET

What Is HYTELNET?

HYTELNET is a tool for identifying and locating Telnet sites on the Internet. HYTELNET provides a menu-driven, hypertext interface, allowing users to browse through menus to find sites of interest, and then make Telnet connections to them.

HYTELNET was developed by Peter Scott at the University of Saskatchewan to serve as a front-end software program for installation on one's own network-connected computer. Versions of HYTELNET software are available for PC, Macintosh, UNIX and VMS systems. Also, there are a few publicly-accessible HYTELNET servers available on the Internet.

Applications That Use HYTELNET

Identify, locate, and connect to Telnet-accessible sites such as:
- Library catalogs
- Indexes, abstracts and bibliographies
- Bulletin board systems
- Campuswide information systems
- Community online services, Free-Nets
- Numeric databases

How to Use HYTELNET

Anonymous FTP Site for Retrieving HYTELNET Software
```
%  ftp ftp.usask.ca
   cd/pub/hytelnet
```

Gopher-based HYTELNET Gateway
```
%  gopher liberty.uc.wlu.edu
        > Explore Internet Resources/
        > Telnet Login to Sites (HYTELNET)/
```

World-Wide Web-based HYTELNET Gateway
```
%  lynx http://www.usask.ca/cgi-bin/hytelnet
```

Exercise Demonstrating Applications of HYTELNET

Literature Exercise 1

For Further Information

For more information, see the HYTELNET Fact Sheet included in *Crossing the Internet Threshold*, 2nd ed. (p.136).

IRC: Internet Relay Chat

What Is IRC?

IRC is a client/server program allowing multiple Internet users to exchange messages simultaneously. IRC sessions take place on particular 'channels', in much the same way CB radio operates. In this way, IRC serves as a real-time conferencing system for the Internet.

IRC has been, traditionally, a tool in favor among recreational or other non-research users of the Internet. But in recent years, a number of science-related IRC channels have been established.

To start using IRC, login, join a channel, choose a nickname, and start chatting. Text that you type is automatically broadcast to all others on that channel. You may also start a new channel, and control its membership by making it 'secret,' 'anonymous,' or 'private.'

You may have an IRC client running on your own Internet node, or may need to use a publicly-accessible IRC client available on several servers around the Internet.

Applications That Use IRC

Online discussion groups in real time

Exercise Demonstrating Applications of IRC

Conferencing Exercise 5

For Further Information

"IRC FAQ Document", "IRC Primer", and other related documentation
> `ftp://cs-ftp.bu.edu/irc/support/`

Jughead

What Is Jughead?

Jughead is an indexing and retrieval program that helps users identify and locate items on selected Gopher menus. Jughead's developer, Rhett "Jonzy" Jones of the University of Utah, appears to have contrived the acronym (Jonzy's Universal Gopher Hierarchy Excavation and Display) to create a companion for Veronica. In fact, Jughead operates much like Veronica, but over a limited portion of Gopherspace. Some Jughead programs are set up to search all the menus of a particular Gopher. Other versions of Jughead, such as the one at Washington & Lee (see example below) search all Gophers around the Internet, but burrow down through only the uppermost levels in the hierarchy.

Jughead is useful when the user reaches the Gopher known to offer a desired resource, but doesn't want to browse through menus to find it. Jughead is also a good alternative to Veronica when access to the latter is unavailable, or when an search term is thought to be so ubiquitous that a limited search will produce sufficient results. Several subject-oriented Jughead services, such as the "Biological Gophers"

> `gopher://gopher.gdb.org:  3005/7`

provide an efficient way of searching Gopherspace for resources for a particular research topic.

As with Veronica, Jughead supports Boolean operators AND, OR and NOT as well as truncation with the asterisk. A successful Jughead search provides a tailor-made Gopher menu of the directories and/or documents matching the user's search terms. The user may then use the '=' key to return the exact location of the items, or may simply add desired menu items to his or her personal bookmark.

Jughead is found on Gopher menus, but it may not always be marked as such. Look for a menu option that reads something like "Search all menus on this Gopher by keyword <?>"

Applications That Use Jughead

- Identify and locate resources of interest on subject-oriented gophers

- Search the top few levels of all gophers around the Internet

- Search for a known resource on a particular gopher

How to Start Jughead

```
%  gopher liberty.uc.wlu.edu 3002
>  Search High Level Gophers by JUGHEAD at W&L <?>
```

For Further Information

"About Jughead" file at the University of Utah

```
gopher://gopher.utah.edu:70/00/

Search%20menu%20titles%20using%20jughead/
about20%jughead
```

"Jughead FAQ" at Washington & Lee
```
gopher://liberty.uc.wlu.edu:70/00/gophers/jughead
```

Mailservers

What Are Mailservers?

Mailservers are programs that automate the delivery of files via electronic mail. Most commonly, mailservers are used to manage electronic discussion groups, but they also serve to distribute newsletters and other text files either by subscription, or on demand from a user.

The family of software known generally as mailservers includes many different programs, each with specific commands and capabilities. Among the most common mailserver programs in use on the Internet are:

> Listserv
>
> Mailbase
>
> Mailserv
>
> Majordomo
>
> Unix Listprocessor

Each of these programs use varying commands and features, but all share a common set of functions:

- Maintain subscriber lists
- Customize subscriber options
- Provide lists of subscribers upon request
- Index and send archive files
- Provide list of commands upon request

For information about specific commands used in the mailserver programs listed above, refer to Jim Milles' helpful guide, "Discussion Lists: Mail Server Commands," referenced below.

You may also encounter other mailservers than those listed above, and each may have different commands and capabilities for requesting information and documents. When you don't know what type of commands to use for a given mailserver, try sending a message with the word 'help' in either the subject line or the message field. Even if the mailserver does not recognize this form of request, it may automatically send a helpful message in response, providing instructions on requesting information and other documents from the mailserver.

There are also various types of Internet discussion groups and mailing lists that are confused with mailservers, but use entirely different methods of operation. For example, Network or Usenet Newsgroups run as services independent of the electronic mail system. For more information about Network/Usenet News, see the corresponding Fact Sheet in this guide.

There are also electronic mailing lists that are administered by real people (not by mailserver software)—these distribution lists take advantage of aliasing capabilities in one's local mail program that 'duplicate' incoming messages and forward them to a list of prescribed electronic mail addresses.[1] However, because the distribution list itself must be compiled and maintained manually, this type of mailing list is best used for very small groups. For more information about the operation of discussion groups of this type, you may need to contact the moderator.

Applications that Use Mailservers

- Join, participate in, and search the archives of electronic discussion groups

- Subscribe to electronic newsletters, bulletins, journals, journal previews, and other publications

- Request documents, texts, and other files on demand

Exercises Demonstrating Applications of Mailservers

Conferencing Exercises 1 and 2
Finding Colleagues Exercises 3 and 4
Reference Exercise 2
Sharing Data Exercise 3

For Further Information

Milles, James, "Discussion Lists: Mail Server Commands"

ftp://ubvm.cc.buffalo.edu/nettrain/mailser.cmd

mailto://listserv@ubvm.cc.buffalo.edu

(Leave subject line blank; in the message field, type: get mailser cmd nettrain f=mail)

Tennant, R., et al. "Electronic Discussions," *Crossing the Internet Threshold*, 2nd ed. Berkeley: Library Solutions Press, 1994, (p. 49-57).

---

1   Usually this procedure requires 'root' permission on the local computer system—see your local system administrator for more information.)

MBone: Multicast Backbone

What Is MBone?

MBone is a virtual network that allows the transmission of live video (one-way) and audio (two-way) over the Internet. MBone runs over portions of the Internet's physical infrastructure, but currently uses its own particular hardware and protocols to route multicast packets between MBone hosts.

MBone uses a number of standards and protocols to multicast video and audio across the Internet: IP Multicast Addressing Standard, the Real-Time Protocol, the Distance Vector Multicast Routing Protocol, and others. Fortunately, one does not need to fully understand these underlying standards and protocols to successfully use MBone. What MBone users do appreciate is the power of real-time, world-wide video- and audio- broadcasting and conferencing, accessible from one's own desktop.

The recent 'broadcast' of a Rolling Stones concert over MBone illustrated to the world at large the potential power of Internet multicasting. In the scientific community, however, MBone has offered numerous innovative applications for research and instruction for several years. Using this technology, the Internet Engineering Task Force (MBone's creators) sends audio and video feeds of their meetings for the benefit of those unable to attend; NASA broadcasts shuttle launches in real-time; and the Naval Postgraduate School Visualization Labrotary has run remote training courses. As noted by naval researchers Macedonia and Brutzman[1] "many of the most exciting events on the Internet appear on MBone first."

Because MBone technology can transmit real-time images from remote instruments or vehicles, it also has significant application in the exploration of space, sea, and other frontiers not directly accessible by humans. One of the most dramatic examples of MBone application is the Jason project from Woods Hole Oceanographic Institution. The remotely-operated submersible/robot Jason/Medea records sonar and other oceanographic data, which is transmitted immediately to modelers onshore. From the raw data, the modelers generate video displaying seafloor features and providing other 'real-time' information that is helfpul to shipboard researchers in analyzing the progress of their work. Video generated by Jason is also broadcast to classrooms via live satellite.

---

1   Macedonia, Michael R. and Donald P. Brutzman. "MBone Provides Audio and Video Across the Internet". *IEEEComputer* vol. 27, no. 4 (April 1994): 30-36.
    Also available on the Internet:
    `file://taurus.cs.nps.navy.mil/pub/mbmg/mbone.html`

Fuller integration of MBone technology in scientific research is hampered by the significant 'costs' involved in transmitting and receiving communications via MBone. You need bandwidth of at least T1 (45 mbs); dedicated routers and other hardware; and an MBone-savvy network engineer willing to set up an MBone host. MBone users need multimedia workstations to produce and receive audio and video transmissions.

Applications That Use MBone

- Real-time transmittal of video or audio data from remote vehicles or instruments

- Real-time video and/or audio-conferencing, with shared "whiteboard"

- Interactive distance learning or training

- Internet broadcasts (*e.g.*, the Internet Talk Radio show, NASA Shuttle launches)

How to Start MBone

You will need to contact your system administrator to find out how/if your site can join MBone. If your site is not connected, it may be possible to connect to a nearby site that is already on the MBone. Refer to the "Frequently Asked Questions (FAQ) on the Multicast Backbone" referenced below for more information.

Once connected to MBone, you can find out what 'events' are running by subscribing to an MBone event mailing list. For example, MBone events in North America and Japan are announced on the lists

> **mbone-na@isi.edu**
> [subscription requests to **mbone-na-request@isi.edu**]

or **mbone-jp@wide.ad.jp**
> [subscription requests to **mbone-jp-request@wide.ad.jp**].

For Further Information

Bunn, Jean, "MBONE (Multicast Backbone)" at Geneva University
> **http://www.unige.ch/seinf/mbone.html**

Casner, Steve. "Frequently Asked Questions (FAQ) on the Multicast Backbone"
> **ftp://venera.isi.edu/mbone/faq.txt**
> **http://www.research.att.com/mbone-faq.html**

"Guide to MBone Etiquette"
> **http://www.eit.com/techinfo/mbone/etiquette.html**

MBONE Information Web (Links to everything related to MBone)
> **http://www.eit.com/techinfo/mbone/mbone.html**

MIME: Multipurpose Internet Mail Extensions

What Is MIME?

MIME is an Internet specification developed to enhance the text-only capabilities of the Simple Mail Trasfer Protocol (SMTP). MIME enables mail messages to include binary data or non-ASCII character sets along with a regular text message, and then transport them using SMTP.

MIME is not a protocol or tool in itself. Rather, MIME adds new fields into the standard mail message header, including a 'MIME-version' field to indicate that a message conforms to MIME standards, and a 'Content-Type' field that specifies the type of data included in the message.

MIME is supported by many Internet mail software packages, including the popular program PINE available on most Unix systems. In order to exchange messages with MIME, both sender and receiver must use mail programs (though not necessarily the same one) that support the MIME specification.

There are also experiments underway to adapt Gopher and World-Wide Web clients to transfer multimedia information using MIME.

Applications That Use MIME

- Deliver manuscripts or other documents in a 'rich' (word processing or Postscript) format
- Exchange email messages in a language that uses non-ASCII characters
- Attach binary files, including sound recordings, images, executable software, datasets, *etc.*, to an email correspondence
- Receive multimedia publications via email

How to Use MIME

A number of popular mail systems support MIME, including PINE and some versions of Eudora. These systems provide mail headers with an ATTACH: field where the user indicates the items to be included with the mail message.

For Further Information

Comp.mail.mime Newsgroup. MIME Frequently Asked
Questions (FAQ)

> `ftp://rtfm.mit.edu/pub/usenet-by-group/`
> `news.answers/mail/mime-faq/`

> `http://www.cis.ohio-state.edu/text/faq/usenet/`
> `mail/mime-faq/top.html`

Grand, Mark. "MIME Overview", October 1993.

> `ftp://ftp.netcom.com/pub/mdg/mime.txt`

> (or `mime.ps`)

MOO: Multiuser Dimension, Object Oriented

What Is MOO?

MOOs are a form of MultiUser Dimension, Object Oriented—a text-based virtual environment on the Internet. Visitors to a MOO navigate through rooms and other sections of a 'building', interacting with other dwellers or with objects found in their wanderings. What's more, users may expand the environment at any time by engineering additional rooms, spaces, tools, or objects for others to view, read, or use.

Adpated from existing Multiuser Dimension (MUD) applicatons at Parc Research Center, Xerox, MOO enables to actively construct their environment, not just play roles within it. Though historically used for game-playing and other recreational pursuits on the network, MOO is now finding application among the scientific community as a system for interactive communication, collaboration, and other directed activities within a meaningful context. The biologists' BioMOO[1] provides a lounge, laboratories, offices, and a seminar room. The astronomers' experimental Astro VR[2] provides a means of running software, retrieving satellite images on demand, searching databases. Colleagues may also show an image while conversing, share a whiteboard, and more.

MOO originated as a text-based system, but researchers are rapidly adapting this tool for full multimedia display and retrieval. The developers of Astro VR, for example, have created specialized client software that allows the user to take full advantage of MOO images, graphics, etc. BioMOO offers a user-friendly World-Wide Web interface in addition to their traditional Telnet service. Other developers are exploring MOO-to-Web and Web-to-MOO applications, finding ways to attach an image, sound, or video file to an object. The promise of MOO/WWW applications, offering the combined power of full multimedia and instant interactivity, may one day become a compelling alternative to the real-life lecture halls, conference rooms, and laboratories of the scientific research center.

MOO is a client/server program. Public clients are available by Telnet, Gopher, and increasingly, the World-Wide Web.

---

1 You may connect to BioMOO at:
    `telnet://bioinfo.weizmann.ac.il:8888`

2 You may connect to AstroVR at:
    `telnet://astrovr.ipac.caltech.edu:8888`

Applications That Use MOO

- Conferencing in real-time

- Journal clubs

- Collaboration on research projects

- Educational programs: online classrooms, electronic experiments, field trips, *etc.*

How to Start MOO

%	`telnet bioinfo.weizmann.ac.il 8888`
connect	**Guest**
Type	**'tutorial'** to learn the basics of MOO tutorial

When ready to leave MOO, type: **@quit**

Exercise Demonstrating MOO

Conferencing Exercise 6

For Further Information

"AstroVR Help Page"

> **http://brando.ipac.caltech.edu:8888/AstroVRHelp**

"InternetVR Sites" Page at Cardiff University, Wales

> **http://www.cm.cf.ac.uk/User/Andrew.Wilson/VR/sites.html**

Mostly recreational, but the scientific-related MOO's are listed at the end under 'Restricted Access/Special Interest'.

MOO Collection of manuals, programs and papers

> **ftp://parcftp.xerox.com/pub/MOO/**

Network News

What Is Network News?

Network News is the Internet's answer to electronic forums available on popular online services or dial-up bulletin boards. On the Internet, however, newsgroups may include thousands of members, participating in heated, scholarly, or highly technical discussions on most every subject imaginable.

The bulk of the newsgroups comprising Network News come from Usenet, a loosely structured distribution system that passes streams of messages, or 'news feeds' to Unix computers worldwide. Originally developed in 1979 by Duke University and the University of North Carolina, Usenet quickly grew to include sites around the world. Usenet now reaches nearly 3 million users at almost 100,000 participating sites. The Usenet News system includes many hundreds of groups, clustered into several topical hierarchies such as 'comp.*' for computer-related interests (*e.g.*, comp.benchmarks, comp.lang.c++, and comp.sys.mac.scitech); or 'sci.*' for science-related interests (*e.g.*, sci.astro.fits, sci.engr.civil, sci.math.num-analysis).

In addition to the Usenet Newsgroups, the Network News system also includes local or regional newsgroups, alternate groups that didn't go through Usenet's group creation process, or proprietary groups that may be fee-based.

Unlike Internet mailserver discussion groups, the Network News system does not distribute postings via electronic mail. Rather, Network News is distributed as a continuous stream of messages, and it is up to the news server at each individual institution to determine which and how many Newsgroups to carry. Most pick up all of the groups in the conventional Usenet hierarchies (*e.g.*, 'comp.*', 'sci.*', etc.), along with select groups in alternate hierarchies (*e.g.*, 'alt.*', 'ieee.*') and any local groups distributed just in that institution or geographic region (*e.g.*, 'fiu.*' or 'fl.*')

Newsgroup users follow the discussions of desired groups by running newsreader software on their local Internet host. Messages posted to the group are mostly text, but binary files containing picture, graphs, and other images may also be available.

Most newsreader commands are comparable to those available with mailserver groups—subscribe, unsubscribe, reply to an individual or to the group, save an article to a file, and so on. But newsreader software also has capabilities for organizing related messages into topical 'threads', deleting or saving entire threads, etc. that facilitate ongoing discussion more effectively than possible in mailserver groups. For that reason, some prefer to use Network News for electronic discussion groups on the Internet.

Applications of Network News

- Communicate and exchange ideas with colleagues with similar interests

- Search newsgroup archives for previous discussions

- Read and download Frequently-Asked-Questions (FAQ) documents about manypopular or 'hot' topics in science, or about Internet sites related to a particular subject area

- Receive electronic newsletters, bulletins, announcements, journal tables of contents, etc. that are routinely posted to various newsgroups

- Identify and download maps, photos, images, sound files, software, *etc.*

How to Start Using Network News (Using Unix tin newsreader)

```
%    tin
         reading news active file...
         reading attributes file...
         reading newsgroups file...
```

You will then see the "Group Selection" list. To subscribe to any group, mark it with an '**s**'.

If you know the group you want to read, you may specify it at startup:

```
%    tin  sci.geo.meteorology
```

Exercises Demonstrating Applications of Network News

Conferencing Exercise 3
Literature Exercise 4

For Further Information

Hahn, Harley, and Rick Stout. *The Internet Complete Reference*. New York: Osborne McGraw Hill, 1994.

Chapters 9-15 provide a very thorough treatment of USENET News and various UNIX newsreaders.

Salzenberg, Chip, "What is USENET?" Available in two parts

```
ftp://rtfm.mit.edu/pub/usenet/news.answers/
usenet/what-is/part1,  part2
```

Telnet: Remote Login

What Is Telnet?

Telnet is one of the original Internet protocols, and it is still in wide use today. Telnet enables users to log in to a remote computer on the Internet and use selected resources: search a database, run a program, control an instrument, etc.

Login via Telnet requires a username and usually a password. During the login process, the remote system may also ask for the type of terminal emulation to use for the session. The full screen VT100 emulation is typically the default because it is so widely used, but a menu of other choices may display.

For Telnet services offered on an IBM mainframe, a special form of Telnet—tn3270—controls proper screen display and functioning of the keyboard. Some Telnet clients automatically switch to tn3270 mode, so no special command is needed. However, if your Telnet session doesn't look or act quite right, if you see garbled characters on the screen or if certain keys don't work, try closing the session and reconnecting with the 'tn3270' command (see below for an example).

Another special form of Telnet is needed for services that use a non-standard port on the remote host. Normally, Telnet users connect through port 23 and don't need to specify the default to make a connection. For services accessible through other ports, however, the exact port number must be specified at the end of the address (see below for an example).

The most problematic part of the Telnet session may be logging off. Each host system has its own command for ending the session and closing the connection (*e.g.,* exit, logoff, logout, quit), and often that command scrolls by in the flash of an eye right after log in. If you didn't catch it then, you may be able to 'look it up' from the online help system—try typing 'H', 'h', or '?' to call up the help options. If nothing works, try using the universal Telnet 'kill' command—control-right bracket or ^]. If you get back to a 'telnet>' prompt, type 'quit' to end the Telnet session.

One of the most useful applications for scientists on the road is Telnetting back to one's own Internet node to check email, perform other work, etc. If you are planning to do this, it is advisable to notify your system administrator of your intentions before you go. Also be sure to verify your full domain name address (usually, what's to the right of the @ in your email address) and numeric IP address, to ensure that the host you will be Telnetting from can find your home machine.

Applications That Use Telnet

- Log in and use your home computer while on the road

- Connect to and search online databank services such as STN, DIALOG, etc.

- Use a publicly-accessible Internet client (*e.g.*, Archie, WAIS, IRC) you don't have on your own computer

- Run a program on a remote computer

- Search bibliographic databases, such as library catalogs & periodical indexes

- Search online reference books and numeric databases

- Operate an instrument remotely

How to Start Telnet

Simple Telnet command

%	**telnet e-math.ams.org**
login:	**e-math**
password:	**e-math**
term:	**vt100**

Telnet to an IBM mainframe

%	**tn3270 nervm.nerdc.ufl.edu**
Command:	**dial vtam**
VTAM ACTIVE	
	nerluis

Telnet to a non-standard port

%	**telnet 141.212.196.177 3000**

No login required.

Exercises Demonstrating Applications of Telnet

Finding Colleagues Exercises 2, 3 and 6

Literature Exercise 1

For Further Information

Tennant, R., et al. "Internet Remote Login (Telnet)", *Crossing the Internet Threshold*, 2nd ed. Berkeley: Library Solutions Press, (1994), 65-73.

URL: Uniform Resource Locator

What is a URL?

The Uniform Resource Locator, or URL, serves as a precise address for a file, directory, searchable database or other service on the Internet. Originally, URLs were developed to specify a particular link made on the World-Wide Web. Now the convention is used more widely to represent the location and access method for most any Internet resource.

URLs act something like library call numbers, pinpointing the location of a resource at a given site, in the same way a call number locates a book on the shelves of a particular library. It is possible that a resource on the Internet may have several URLs—each pointing to a different location for that same file or service.

The URL follows the generic syntax "scheme://scheme-specific-part". The "scheme" represents the particular protocol or access method for the resource; the "scheme-specific-part" details the host, port, and path or other access information specific to that scheme.

URLs exist for many types of Internet services, including FTP, http (WWW), Gopher, mailto (electronic mail), USENET news, Telnet, and WAIS. Descriptions and sample URLs for some common resources follow.

FTP URL

The FTP URL designates files and directories accessible with the FTP protocol. FTP URLs allow specification of a username, password, and port. If none are supplied, anonymous FTP is assumed through default port 21. If the host requires an email address as a password, the user will be prompted for one.

Sample URL for anonymous FTP:

```
ftp://stis.nsf.gov/fm1207.ps
```

Telnet URL

Telnet URLs represent an interactive service. They allow specification of a username, password, and port. If none are supplied, the port defaults to 23, and the remote host may require the user to input a login name, password, and emulation type.

Sample URL for telnet

```
telnet://stis.nsf.gov/
```

Gopher URL

Gopher URLs accommodate all types of Gopher menu items — directories, files, and even searchable databases and other interactive services. If not specified, the port defaults to 70.

The scheme-specific-part of the Gopher URL details the complete path to the resource or menu item, beginning with the Gopher item type (*e.g.*, '0' for file, '1' for directory), which, for some unknown reason, occurs twice consecutively. For an explanation of Gopher item types, see the Gopher Fact Sheet elsewhere in this section.

An especially cryptic part of the Gopher URL are the escape characters inserted to designate spaces in a directory title or filename. The code '%20' between words, as in the gopher URL

gopher://gopher.tc.umn.edu/11/Other20%Gopher20%and...

represents the spaces in the directory title "Other Gopher and ..."

Sample URLs for Gopher:

gopher://gopher.lib.ncsu.edu/11/library/

(Gopher directory;type=1)

gopher://x.nsf.gov/00/NSF/forms/fm1207.ps

(Gopher text file;type=0)

HTTP URL

HTTP URLs designate resources accessible with the Hypertext Transfer Protocol (HTTP) on the World Wide Web, including hypermedia documents, files, and even searchable indexes. The port defaults to 80 if omitted.

Sample URL for HTTP

http://stis.nsf.gov/nsf/pubs.htm

Use of URLs in This Guide

For the sake of uniformity, nearly all Internet addresses provided in this guide follow URL syntax. That does not mean, however, that users must invoke a Web browser to access the sites listed in this guide (with the exception of resources served only by the HTTP protocol.) One may use whatever protocol is designated in the 'scheme' portion of the URL. For example, the file "Introduction to the Directory of Electronic Journals" represented by the URL:

gopher://arl.cni.org:70/00/scomm/edir/eintro

may be accessed directly from a Gopher client as follows:
```
% gopher arl.cni.org
        - >Scholarly Communication/
            - > Directory of Electronic Journals/
                - >Introduction.
```

For Further Information

Andreessen, Marc. "A Beginner's Guide to URLs"

`http://www.ncsa.uiuc.edu/demoweb/url-primer.txt`

A good place to start.

Berners-Lee, T. "Universal Resource Identifiers in WWW: A Unifying Syntax for the Expression of Names and Addresses of Objects on the Network as Used in the World-Wide Web" *RFC 1630*, June 1994.

`gopher://nic.merit.edu:7043/00/internet/documents/`
`rfc/rfc1630.txt`

Probably the most complete, authoritative document explaining the syntax and interpretation of Uniform Resource Identifiers (URI), of which the URL is one specific type.

Berners-Lee, T., Masinter, L., and McCahill, M. "Uniform Resource Locators (URL)", Internet Draft (Work In Progress) of the IETF URI Working Group.

`gopher://nic.merit.edu:7043/00/internet/documents/`
`internet-drafts/draft-ietf-uri-url-08.txt`

This working paper is a continually updated draft recording the current use and development of URL.

Morgan, Eric Lease. "The World-Wide Web and Mosaic: An Overview for Librarians." The Public-Access Computer Systems Review 5, no. 6 (1994): 5-26.

`gopher://info.lib.uh.edu:70/00/articles/e-journals/`
`uhlibrary/pacsreview/v5/n6/morgan.5n6`

This article explains in clear and simple terms how to interpret various URL types.

Veronica

What Is Veronica?

Veronica is an indexing and retrieval system that helps users locate any item on a Gopher menu most anywhere on the Internet. Developed by Steve Foster and Fred Barrie at the University of Nevada, Veronica was designed to build indexes of Gopher menus in the same manner that Archie indexes files on FTP servers.

Veronica is found on most Gopher menus. When selected, the Veronica menu will display a list of the Veronica servers worldwide—at present, there are about eight. All Veronica servers offer essentially the same index, although small differences occur due the exact timing of the last indexing run.

Veronica offers two types of searches: searches of directory titles only (that is, Gopher menu items that represent the top of a hierarchy); or searches of <u>all</u> items appearing in Gopherspace (that is, every file, every link to a telnet site, etc.) Usually a search on directory titles is sufficient to find key sites relevant to the user's query.

Veronica also offers several search features: the boolean operators AND, OR, and NOT; truncation with an asterisk (*), and search limits by Gopher item type (for a list of these types, see the Gopher Fact Sheet).

A successful Veronica search provides a tailor-made Gopher menu of all directories and/or documents matching the user's search terms. Users may then work with this menu just like any other Gopher menu, browsing, burrowing, saving files, etc. The biggest difference in the Veronica menu is its ephemeral nature—upon exiting from the menu, it dissolves.

Because Veronica-compiled menus are not permanent, users need to know the Gopher options for bookmarking links the user wishes to return to, or saving desired documents for later use. One may add any item to a personal bookmark file with the 'a' key, or determine the exact location of the item with the '=' key. Also, by adding the '- l' flag to any Veronica query (see the example below), some Veronica servers save the retrieved links to a 'link info file', provided as the first item of the results menu.

Finally, keep in mind that the Veronica servers are heavily-used and at times may be difficult to access. If you are unable to access Veronica, you might try a server residing in another time zone. Alternatively, you may use Jughead to conduct a search of Gopherspace — the scope of the Jughead search is not as comprehensive as a Veronica search, but may still turn up enough Gopher sites to meet your need. For more information on Jughead, see the corresponding Fact Sheet.

Applications That Use Veronica

- Identify and locate any resource of interest in Gopherspace

How to Start Veronica

%   **gopher gopher.unr.edu**

- > Search ALL of Gopherspace (4800 servers) using Veronica/
- > Find GOPHER DIRECTORIES by Title word(s)
      (via Univ...)<?>

Words to search for : **mycolog* -l**

Veronica compiles the resulting menu:

Find GOPHER DIRECTORIES by Title word(s) (via University of Koeln): mycolog* -l

- > 1. *** Link info for mycolog* ***.
  2. Mycologists Online (MYCO)/
  3. Regional Mycology Reference Laboratory (organizationalUnit)/
  4. Mycological Society of America Bulletin Board/
  5. Frequently Asked Questions About Mycology/
  6. mycology/
  7. mycology/
  8. Botany and Mycology/
  9. bionet.mycology/
  10. Mycological Research/
  .   <display deleted>

Exercises Demonstrating Applications of Veronica

News Exercise 3
Reference Exercise 4
Literature Exercise 2

For Further Information

"Veronica FAQ"

**gopher://veronica.scs.unr.edu/00/veronica/
veronica-faq**

Foster, Steven. "How to Compose Veronica Queries"

**gopher://veronica.scs.unr.edu/00/veronica/how-to-
query-veronica**

WAIS: Wide Area Information Service

What Is WAIS?

WAIS is a finding tool used to search and retrieve indexed documents on the Internet. It is based on the Z39.50 protocol developed in the library/information science community to enhance information searching and retrieval.

WAIS indexes cover the contents of a particular file, not just its filename or directory title (in contrast to the tools Archie and Veronica). And because the indexes are often built from every word in the document, WAIS searches are generally "full text."

WAIS is a client-server application. Users enter their search terms from a client interface running on their own systems, or from one of the publicly-accessible WAIS clients (see below). Each client sends queries to and returns results from one of the many WAIS servers on the Internet.

Each document collection that WAIS searches is termed a WAIS 'source', denoted with an .src extension. There are now over 500 sources available throughout the Internet. Most tend to focus on a particular subject and type of resource, such as job listings for astronomers (aas-jobs.src), descriptions of pictures at the Smithsonian (smithsonian-pictures.src), or genome data (genethon_seq.src).

The WAIS client searches for user's keywords in one or more selected sources and returns a list of documents ranked by relevancy. Documents containing the greatest number of occurrences of the search term(s) appear on top, with a maximum relevance ranking of 1000.

The searcher may then browse through each retrieved document, selecting desired documents to save or to serve as a basis for "finding more of the same." This latter feature of WAIS is termed 'relevance feedback.'

The search options and capabilities of WAIS vary with the version of WAIS in use. Three versions of WAIS are currently found on the Internet. Simple WAIS, or Swais, is the original version developed by Brewster Kahle[1] and it provides for simple keyword searches, an implied 'or' connecting multiple terms, and truncation with an asterisk. Both IU Bio WAIS, developed by Indiana University, and freeWAIS from the Clearinghouse for Networked Information Retrieval (CNIDR), do allow boolean searches including ' and,' 'or' and 'not.'

One may run a WAIS search from a publicly-accessible client (as shown below) or by running a WAIS client on one's own computer. The latter is preferable because it allows searchers to use certain source files not accessible from the public clients. Also, if a graphics-capable WAIS client is running, such as MacWAIS or WinWAIS, one may display images retrieved during a WAIS search. However, those using Telnet to a publicly-accessible WAIS client will be able to perform most tasks requiring WAIS.

Users of other Internet tools may also run WAIS unknowingly. Veronica, Jughead and most WWW services use WAIS to locate and retrieve items matching the user's keywords.

Resources that may be searched and retrieved using WAIS

- Bibliographies and indexes
- Collections of full-text documents such as journal articles, technical reports,etc.
- Archives of Usenet News and other discussion groups
- Collections of images
- Collections of data
- Directories of people
- Reference texts or databases

Exercises Demonstrating Applications of WAIS

Finding Colleagues Exercise 2
News Exercise 4
ReferenceExercise 5
Literature Exercises 3 and 6

For Further Information

"WAIS FAQ Document"

```
ftp://rtfm.mit.edu/pub/usenet/news.answers/wais-
faq/getting-started
```

"WAIS Bibliography"

```
ftp://ftp.cnidr.org/pub/wais-wais-discussion/
bibliography.txt
```

---

1   At the time he developed WAIS, Kahle worked for Thinking Machines and then Apple Computer.

World-Wide Web (WWW) Robots

What Are WWW Robots?

The general term 'WWW robots' refers to a class of specific tools for discovering resources on the World-Wide Web. These tools are also known variously as spiders, crawlers, wanderers, worms, and other creeping species. Examples of these tools are:

> Harvest
>
> JumpStation
>
> WebCrawler
>
> WWWW: World-Wide Web Worm
>
> Lycos
>
> Peregrinator

What distinguishes WWW robots from other resource discovery tools for the World-Wide Web is their automated nature: each of these tools use software agents that are programmed to traverse the Web on a regular basis to scout for new resources, and include them in a knowledge base of resources and subjects. Other Web search tools may also provide the capability of searching the Web, but draw on a knowledge base that is compiled manually, through volunteer submission, manual screening and compiling, or other means.

How Do They Work?

The term 'tool' is a bit of a misnomer because most of the robots referenced above are actually a set of tools that work together to provide an automated resource discovery system for the Web. These systems generally include three components:

(1) The wanderer

This component traverses the Web, finding and extracting new resources for entry into the knowledge base. Each robot differs in the frequency of its search (daily, every few days, etc.). and in the scope of its traverses. Some robots may limit their travels to a specific range of servers; others may traverse any server they can find, but follow only a small number of links from the 'top' page. Some robots limit searching to documents served by the HTTP protocol, whereas others may also traverse information served by Gopher, FTP, *etc.*

(2) The index or knowledge base

The information extracted from the WWW documents traversed by the robot is compiled into an index or knowledge base containing locations of resources (specified by a URL) and some context /descriptive information about the resource. The type and extent of included information differs from one robot to the next. Some robots index the full text of the Web document; others include only a small window of text surrounding each link. Still others index certain sections of the Web page, such as the header or the title.

The size of the indexes also differs substantially. The Lycos robot, for example, now references a knowledge base of over 2.5 million URLs. It is also one of the slower systems to search. Keep in mind that size of the index is not necessarily related to the effectiveness of its coverage—some robots, such as Harvest, have developed more efficient mechanisms for providing comprehensive coverage of the Web without massive storage of indexed information.

(3) The search system

The search system includes both the user interface, where users enter their search terms, and the search engine that finds material of interest to the user. Most robots provide an interactive HTML form which requires the user to have a graphical Web browser. But many robots also provide an alternate search interface for those using Lynx or another non-graphical interface.

The search engines in use by the various WWW robots vary in their capabilities. Possible options include, among others, case sensitivity/insensitivity; Boolean connectors *and, or, not*; partial word matching; phrase matching; wildcards. Some allow you to limit matches to a particular field or to a specified number of hits, or to specify results to be sorted or ranked by relevance.

You cannot count on retrieving the same information from different robots because of their different indexing algorithms. For this reason, it is a good idea to try the same search using more than one robot.

The specific features and capabilities of many WWW robots (and other WWW search tools) are discussed in Section C of this guide. They are also detailed in the "World Wide Web Wanderers, Spiders, and Robots" page, referenced below.

Exercise that Demonstrates the Use of WWW Robots

Reference Exercise 3

For Further Information

NEXOR's "World Wide Web Wanderers, Spiders, and Robots" page

> **http://web.nexor.co.uk/users/mak/doc/robots/robots.html**

NEXOR's well-maintained robots page provides descriptions of all known robots, with links to the home page and search interface for each specific robot.

World-Wide Web: WWW

What Is WWW?

The World-Wide Web is a multimedia information system that integrates global Internet sources into single hypertext documents available for online display or searching. Web documents (also referred to as 'pages') contain text, optional graphics, and hyperlinks—hot areas of the display that, when selected, call up a related file or service somewhere on the Internet. Linked information sources may include another hypertext document, or an image, audio file, video clip, interactive service, or some other type of network file.

The Web was developed in 1990 as a prototype information management system/resource sharing tool for scientists at CERN, the Swiss particle physics laboratory.[1] But the original concept of the Web may be traced back nearly fifty years, to another scientist concerned with the ineffective access to the rapidly expanding scientific record. Vanevar Bush, then director of the national Office of Scientific Research and Development, envisioned a knowledge management system in which related information from disparate sources could be dynamically linked through a "web of trails."[2] The tool he mentally constructed—the memex—implemented the associative selection process natural to human thought. The "private file and library" he envisioned would be mechanized for quick and easy consultation from one's desktop, and would provide useful trails that may be recalled for later use. Much of Bush's prescient vision has now been realized in the today's World-Wide Web.

The World-Wide Web has quickly caught on throughout the research and educational communities, and in the marketplace as well. WWW is now the fastest growing service on the Internet, moving over 1.4 terabytes of traffic as of August 1994, up 700% from the year's beginning.[3]

Today's Web consists of three essential elements that distinguish it from other client/server applications on the Internet:

(1) the underlying Hypertext Transfer Protocol, capable of delivering files from nearly every type of server on the Internet, including FTP, Telnet, Network News, and Gopher, in addition to hypertext documents unique to WWW servers

---

1 Berners-Lee, Tim, Ari Luotonen, Henrik Frystyk Nielsen, and Arthur Secret, "The World-Wide Web." *Communications of the ACM* vol. 37, no. 8 (1994): 76-82.

2 Bush, Vanevar. "As We May Think." *The Atlantic Monthly*: 176 (July1945): 101-108

3 These statistics represent traffic over the NSFNet. Current figures are available from Merit Network Information Center <ftp://ftp.merit.edu/statistics/nsfnet>.

(2)   the Uniform Resource Locator, or URL, currently used to provide precise access instructions of most any resource on the Internet

(3)   the Hypertext Markup Language, or HTML, used to format the hypertext documents presented to the Web user.

Users of resources on the WWW need to gain a basic understanding of the latter two elements—the URL and HTML—in order to successfully retrieve and display information. Both are discussed further in the corresponding Fact Sheets included in Section B of this guide. A fuller understanding of the first element—the HTTP protocol—is important for those wishing to provide or publish information on the World-Wide Web. Further information about HTTP may be found at the information site at CERN, referenced below.

To browse, display, or retrieve information on the WWW requires the use of a WWW client, or browser. A number of clients are in common use on the Internet, offering varying degrees of functionality. At the lower end are text-only browsers that do not offer the capability of viewing graphics and other non-text information online. They do they do allow the user to follow any links that are available in a document, and download non-text files for use offline. Examples of text-only Web clients include the line mode browser developed by CERN, and the full-screen, text-only Lynx, developed by the University of Kansas.

At the higher end of the Web browsers are those programs offering graphical user interfaces, or GUI (pronounced gooey), allowing the user to display or play online any pictures, audio, video, *etc.* that accompany the text. Programs such as Netscape, Mosaic, Cello, MacWeb, and Arena are examples of graphical Web browsers. To run these programs, users must have a graphics-capable computer, such as an XWindows terminal, a PC running Windows, or a Macintosh, and that computer must be directly connected to the Internet, through an ethernet, serial (SLIP, PPP), or other direct connection. They may also need various 'helper applications', software that runs during a WWW session and allows the user to view images, play video or audio clips, display text in special formats such as Postscript, LaTex, *etc.* Information about the various Web clients and helper applications may be found at the CERN site referenced below.

As the Web continues to grow exponentially, and the availability of easy-to-use client software enables more users to access its myriad resources, the most difficult aspect of Web use may be finding information to match a particular need. Fortunately, a number of recently-developed WWW catalogs, directories, and search tools enable searchers to navigate this ever-expanding information space. Many of these resources are discussed in Section C of this guide. The automated tools for Web searching are also discussed in the "Web Robots' Fact Sheet.

Applications of WWW

- Campus-wide (or institutional) information systems
- Online hypertext publications: scientific journals, hyper-reference books and texts
- Single interface to other Internet applications: Telnet, FTP, Gopher, WAIS, MOO
- Hypermedia tutorials
- Library user interfaces
- Software, videos, simulations, etc. that may be played online
- Interactive services, clickable maps, and other images

How to Start WWW (Using the Public Client Lynx)

%	`telnet sunsite.unc.edu`
login:	`lynx`
TERM=	`vt100`

Exercises Demonstrating Applications of WWW

Conferencing Exercises 3 and 4	Literature Exercise 5
News Exercise 6	Publishing Exercises 5 and 7
Reference Exercise 6	Sharing Data Exercises 2 and 4

For Further Information

Berners-Lee,Tim. "World Wide Web Initiative" Geneva: CERN, 1994.

> `http://info.cern.ch/hypertext/WWW/TheProject.html`

Torkington, Nathan. "WWW Primer" (for beginners)

> `http://www.vuw.ac.nz/non-local/gnat/www-primer.html`

"World Wide Web Frequently Asked Questions"

`http://sunsite.unc.edu/boutell/faq/www_faq.html`
(hypertext format)

> `ftp://rtfm.mit.edu; pub/usenet/news.answers/www/faq/part1,part2`

"World-Wide Web/Mosaic Electronic Tutorial" at the USGS

> `http://info.er.usgs.gov:4444/train`

Section C: How to Find Internet Resources In Your Area of Interest

Subject Internet Guides

Subject Hubs

Function Hubs

Subject-Organized Resources

Resource Discovery Tools

Prospecting for Pointers

Section C: How to Find Internet Resources In Your Area of Interest

Finding Internet resources in your own area takes persistence, trial and error, and a bit of luck. The following strategies will help make order out of chaos.

SUBJECT INTERNET GUIDES

Prepared by subject specialists and freely available to all, subject-oriented guides to Internet resources are a great place to start finding sites in your area of interest.

These guides tend to be quite comprehensive, detailing all variety of Internet sites: societies, discussion groups, professional directories, software and data archives, bibliographies, and much more.

The following well-known subject guides are archived at various sites on the Internet.

Anderson, Bob (National Agriculture Library) "Agriculture Based Services and Products - Available on the Internet"

```
gopher://gopher.nalusda.gov:70/00/nalpub/absp_int
```

Briggs-Erickson, Carol and Toni Murphy "A Guide to Envronmental Resources on the Internet"

```
http://www.cfn.cs.dal.ca/Environment/EAC/briggs-murphy-
toc.html
```

Drew, Wilfred. "Not Just Cows: Internet/BITNET Resources in Agriculture and Related Sciences"

```
ftp://info.monash.edu.au/pub/library/guides/agriculture.txt
http://www.monash.edu.au/library/guides/agriculture.html
```

Malet, Gary and Lee Hancock "Guide to On-line Medical Resources", v. 2.3, 3/95

```
gopher://una.hh.lib.umich.edu:70/00/inetdirsstacks/
medclin:malet
http://kuhttp.cc.ukans.edu/cwis/units/medcntr/Lee/
HOMEPAGE.HTML
```

O'Haver, T.C. "Internet Resources for Mathematics and Science Education"

```
gopher://riceinfo.rice.edu:1170/00/Profdevelop/Resources/
resources
gopher://gopher.inform.umd.edu:70/00/Computing_Resources/
NetInfo/ReadingRoom/InternetResources/math-science-edu
```

Smith, Una R. "A Biologist's Guide to Internet Resources" November 1993

```
ftp://rtfm.mit.edu/pub/usenet/news.answers/biology/guide/
part1...part 6
```

Stern, Illana "Sources of Meteorological Data Frequently Asked Questions (FAQ)" (continuously updated)

```
ftp://rtfm.mit.edu/pub/usenet/news.answers/weather/data/
part1, part 2
```

Stevanus, Mary "A Guide to Federal Laboratory and Technology Related Resources via the Internet"

```
gopher://sarah.nalusda.gov:70/00/infocntr/ttic/t2guide.txt
```

Thoen, Bill "On-line Resources for Earth Scientists", June 15, 1994

```
ftp://ftp.csn.org/COGS/ores.txt
```

Wiggins, Gary "Some Chemistry Resources on the Internet"

```
gopher://lib-gopher.lib.indiana.edu:7050/0/chem-data/
resource.ii
```

Clearinghouse For Subject-Oriented Internet Guides

```
gopher://una.hh.lib.umich.edu/11/inetdirs/sciences
```
```
http://www.lib.umich.edu/chhome.html
```

Another source of subject-oriented Internet guides is this clearinghouse, maintained by the University of Michigan's School of Information and Library Studies. Some of the guides included here cover sci-tech topics (e.g., 'Aeronautical Engineering'). They tend to emphasize more traditional library resources (bibliographic databases, library catalogs, etc.)

Most of the guides included at this site resulted from the efforts of graduate students in the School who, as part of a class assignment, have developed subject-oriented guides to the Internet. Since the course runs regularly, a new crop of guides appears from time to time. Some of the guides have not been updated since originally written, but they are very helpful. Also, a few other subject-guides are also deposited in this site, including the various sections of the Kovacs' guide to scholarly electronic discussion groups.

Frequently Asked Questions (FAQ) Documents

Another source of Internet resource listings are the 'Frequently Asked Questions' Documents, or FAQs, compiled by many science-related Usenet Newsgroups to provide commonly-sought information about the subject. Among the questions typically covered in FAQs are "Where are the good Internet sites for this field?" or "Where are relevant data archives, and how do I access them?".

FAQs are authored by moderators or members of the discussion group—hence they are written by working scientists, for their own colleagues. Updated versions are posted to not only the corresponding discussion group, but to other related groups, as well as archived at various FTP, Gopher, and WWW sites.

Some science-related FAQs (and their producers or editors) include:

AIDS FAQ, Dan Greening

Biological Information Theory FAQ, Thomas Schneider, *et al.*

Nonlinear Programming FAQ, John Gregory

Ozone depletion FAQ, Robert Parson

Sea level FAQ, Robert Grumbine

Sci.engr.semiconductors FAQ, Fred Bertsch

Sci.math FAQ, Alex Lopez-Ortiz

Sci.physics FAQ, Scott Chase

Scientific Data Format Information FAQ, Ilana Stern

Sources of Meteorological Data FAQ, Ilana Stern

... and many more!

To see what FAQs are available in your area of science or technology, or to get the latest version of an FAQ, try the following Usenet FAQ archive sites:

```
Scifaq-l@yalevm.bitnet (Science FAQ List)
    mail to      listserv@yalevm.bitnet
    or           listserv@yalevm.cis.yale.edu

    message:     subscribe scifaq-l YourName
```

Subscribe to this mailing list and you will automatically receive the latest version of many of the science-related FAQs.

USENET FAQ Archives at MIT

`ftp://rtfm.mit.edu/pub/usenet-by-group/sci.answers/`

This is the most complete and up-to-date archive site for all sci-tech USENET newsgroups.

MIT's USENET FAQ Collection on Gopher

`gopher://otax.tky.hut.fi/11/English/Topics/FAQs/`

This is a gopherized version of the FTP archives maintained at MIT.

List of USENET FAQ's on WWW

`http://www.cis.ohio-state.edu/hypertext/faq/usenet/`

This hypertext document provides links to the USENET FAQs maintained at MIT, organized alphabetically by the title of the FAQ document (not necessarily the group producing the FAQ). Limited keyword search capabilities are also under development at this site.

SUBJECT HUBS

In many fields of science and technology, one or more network sites offer collections of important resources for a particular subject, with extensive pointers to other relevant sites elsewhere on the Internet. These 'subject hubs' are not officially sanctioned by some authority, and they do not seem to nucleate or develop in any particular pattern. They just seem to appear and then evolve. Often they are maintained by members of an academic department or research institution on a voluntary basis. Because they are developed by researchers in the same field, they tend to be well-selected, offering resources deemed most important or useful by one's own colleagues.

You can think of subject hubs as special libraries or special collections on the Internet. These collections may not be 100 per cent complete, but most are quite extensive, offering 'one-stop shopping' for many different kinds of applications within a given subject area (*e.g.*, discussion group archives, email directories, electronic journals, data archives, and more) They may also contain sub-directories for the diverse sub-specialties with the broader subject.

On the next page are listed a few select subject hubs for some scientific and technical fields.

Astronomy

AstroWeb: Astronomy/Astrophysics on the Internet

    `http://fits.cv.nrao.edu/www/astronomy.html`

    `http://stsci.edu/net-resources.html` [sorted by protocol]

Biology

ANU Bioinformatics

    `http://life.anu.edu.au:80`

Harvard BioPages

    `http://golgi.harvard.edu/biopages.html`

Johns Hopkins BioInformatics Web Server

    `http://www.gdb.org/hopkins.html`

Global Biological Information Servers by Topic

    `gopher://genome-gopher.stanford.edu/11/topic`

Jughead Search Interface to all Biological Gophers

    `gopher://gopher.gdb.org:3005/7`

Chemistry

Yale Chemistry Gopher

    `gopher://yaleinfo.yale.edu:7000/11/Chemistry`

UM-St. Louis Chemistry Gopher

    `gopher://umslvma.umsl.edu:70/11/library/subjects//chmistry`

Chempointers at UCLA

    `http://www.chem.ucla.edu/chempointers.html`

Computer Science

WWW Virtual Library - Computer Science

    `http://ai.iit.nrc.ca/shadow/computing_overview.html`

Earth Sciences

Geology or Geophysics at U. Calgary

    `http://www.geo.ucalgary.ca/VL-EarthSciences.html`

    `http://www-crewes.geo.ucalgary.ca/VL-Geophysics.html`

Michigan Tech's Volcanoes Site

http://www.geo.mtu.edu/volcanoes

German Geosciences Catalogue

http://www.rz.uni-karlsruhe.de/Outerspace/
VirtualLibrary/55.en.html

USGS Directory of Earth and Environmental Sciences

http://info.er.usgs.gov/network/science/earth/index.html

Engineering

ICE: Internet Connections for Engineering (Cornell's Engineering Library)

http://www.englib.cornell.edu

WWW Virtual Library - Engineering

http://epims1.gsfc.nasa.gov/engineering/engineering.html

CAD Centre Engineering Information Guide

http://www.cad.strath.ac.uk/EngInfoGuide.html

Environmental Science

University of Virginia's EcoGopher

gopher://ecosys.drdr.virginia.edu

WWW Virtual Library - Environment

http://ecosys.drdr.virginia.edu/Environment.html

Mathematics and Statistics

Center for Scientific Computing (Finland)

http://www.csc.fi/math_topics/General.html

Florida State Department of Mathematics' Server

http://euclid.math.fsu.edu/Science/math.html

Statistics Education Information Service

gopher://jse.stat.ncsu.edu 70

University of Florida's Statistics Departments' Server

http://www.stat.ufl.edu/vlib/statistics.html

University of Tennessee Mathematics Resources

```
http://archives.math.utk.edu:80/
```

Oceanography/Meteorology

Virtual Library - Oceanography

```
http://www.met.fu-berlin.de/DataSources/MetIndex.html
```

Brookhaven National Laboratory Server

```
http://bnloc7.das.bnl.gov/ocean/index.html
```

Physics

LANL Physics Servers

```
gopher://mentor.lanl.gov:70/11
http://mentor.lanl.gov
```

FUNCTION HUBS

Collections of sci/tech resources related by function, not subject, are also useful tools for finding Internet sites of interest. As with subject hubs, function hubs tend to be maintained voluntarily by an academic department, research institution, or library.

Here are some function hubs that are useful for many scientific and technical fields.

Electronic Journals

CIC Electronic Journals Collection

```
gopher://gopher.cic.net:2000/11/e-serials/managed
```

Government Gophers

Gopher Jewels Listing of Government Gophers

```
gopher:/cwis.usc.edu11Other_Gophers_and_Information_
Resources/Gopher-Jewels/government
```

Government Gophers Collection

```
gopher://peg.cwis.uci.edu:7000/11/gopher.welcome/peg/
GOPHERS/gov
```

WWW Servers (U.S. Federal Government)

```
http://www/fie.com/www/us_govt.html
```

U.S. Government Labs

```
http://boris.qub.ac.uk/edward/GovtLabsUS.html
```

Grants and Funding

Gopher Jewels Listing of Grants Collections

```
gopher://cwis.usc.edu/11/Other_Gophers_and_Information_
Resources/Gopher-Jewels/research/grants
```

Grants Directory (at EINET)

```
http://galaxy.einet.net/galaxy/Reference-and-Interdiscipli-
nary-Information/Grants.html
```

"A Grant Getter's Guide to the Internet"

```
gopher://gopher.uidaho.edu/00/e-pubs/grant
```

University of California, San Diego Science and Engineering Library's
Directory of Grants and Funding Sources

```
http://scilib.ucsd.edu/subjectdir/grants.html
```

Job Opportunities

University of California, San Diego Science and Engineering Library's
List of Employment Resources

```
http://scilib.ucsd.edu/subjectdir/employment.html
```

Patents

Questel/Orbit Patent and Trademark WWW Server

```
http://www.questel.orbit.com/patents/
```

Source Translation & Organization's (STO) Patent Search System

```
http://sunsite.unc.edu/patents/intropat.html
```

University of California, San Diego Science and Engineering Library's
Directory of Patent-related Resources

```
http://scilib.ucsd.edu/subjectdir/patents.html
```

Scientific Visualizations

NASA's Visualizations Annotated Bibliography

```
http://www.nas.nasa.gov/NAS/visualization/visWeblets.html
```

Societies and Associations

Scholarly Societies (via the University of Waterloo)

```
gopher://watserv2.uwaterloo.ca:70/00/servers/campus/
scholars
```

```
http://www.lib.uwaterloo.ca/society/overview.html
```

Standards/Specifications

Document Center

```
http://doccenter.com/doccenter/home.html
```

University of California, San Diego Science and Engineering Library's Directory of Standards-related Resources

```
http://scilib.ucsd.edu/subjectdir/standards.html
```

TeX/LaTex

Comprehensive TeX Archive Network (CTAN)

```
ftp://ftp.shsu.edu/tex-archive
ftp://ftp.tex.ac.uk/tex-archive
```

"Information About TeX" University College, Cork, Ireland

```
http://curia.ucc.ie/info/TeX/menu.html
```

TeX Info Gateway (Large collection of TeX documents, user guides, etc.)

```
http://www.cis.ohio-state.edu/htbin/info/dir
```

SUBJECT-ORGANIZED RESOURCES

If you haven't found a subject hub for your area of interest, the next best thing may be a directory of Internet resources that is organized by subject. Such directories may cover only certain types of applications (*e.g.*, the *Directory of Electronic Journals or Newsletters*; the *Directory of Scholarly Discussion Groups*; Gopher Jewels) or may include a selective list of diverse applications for each subject area (*e.g.*, the Clearinghouse for Subject-Oriented Internet Subject Guides).

The confusing or problematic aspect of subject-organized resources is the criteria by which resources are selected and how they organized. What resources are included or omitted, and how do the developers of the resource decide? What kind of classification system do they use to group resources?

If you are a member of the library profession, or perhaps a regular user of libraries, you may prefer to trust in the subject-organized resources produced or sponsored by a library. A number of such resources are listed below. However, a number of other subject-organized resources not related to libraries are also effective and worthy of attention. These are also included on the list below.

What's most important to keep in mind when browsing through subject-organized resources is that one doesn't need to find an 'all-inclusive' site to begin locating Internet resources in your area of interest. While some of the resources listed below will indeed lead you to a subject hub for your field, others may point to few key resources in your area, and those will inevitably lead you to others. In fact, one of the most 'productive' resources to start finding subject-oriented resources are discussion groups that focus on your area of interest—they frequently post new links of interest to the group. Searching the archives of these lists may also turn up sites you might have missed.

Here is a selective listing of subject-organized Internet resources with substantial coverage of science and technology.

ANU's Electronic Library Information Service (ELISA) using Library of Congress Classification

> `gopher://info.anu.edu.au:70/11/elibrary/lc`

This library site organizes resources by Library of Congress classification. Check out 'G' for geography or oceanography; 'Q' for natural and physical sciences; and 'T' for engineering and technology.

List of Gopher Subject Trees

> `gopher://burrow.cl.msu.edu/11/internet/subject`
> `gopher://munin.ub2.lu.se/11/resources/bysubject`

These sites contain a long list of links to other popular and well-known Gopher subject trees.

Clearinghouse For Subject-Oriented Internet Guides

> `gopher://una.hh.lib.umich.edu/11/inetdirs/sciences`
> `http://www.lib.umich.edu/chhome.html`

See the description in the 'Subject Guides' section above.

Directory of Scholarly Discussion Groups

```
gopher://lib-gopher.lib.indiana.edu/11/research-aid/
acadlist
gopher://gopher.usask.ca/77/.index/acad/
gopher://arl.cni.org/11/scomm/edir/edir94
wais://munin.ub2.lu.se:210/academic_email_conf
```

Compiled by Diane Kovacs and her team at the Kent State University Library, this directory lists and describes discussion groups operated on USENET, Listserv, and other mailserv software in most every research field, with complete subscription directions. The Gopher version of the directory exists in several files, divided by broad subject. The index file helps users identify which groups(s) may be of interest and in which file they will find the relevant description(s).

EELS: Engineering Electronic Library, Swedish University of Technology

```
http://www.ub2.lu.se/eel/eelhome.html
```

Developed and maintained by the Library at SUT, this site selects and organizes only those Internet resources deemed to be high-quality by their team of selectors. Their subject tree follows the classification scheme from Engineering Information, and currently emphasizes resources in engineering (civil, mechanical, electrical, environmental), mathematics, physics, and engineering management. Areas outside the engineering classification include polar research and cold engineering, human work science, and analytical, theoretical, and drug chemistry.

EINET Galaxy Home Page

```
http://galaxy.einet.net/galaxy.html
```

EINET's Home Page is the default starting point for those accessing the World-Wide Web via the MAC client, MacWeb, or the Windows client, WinWeb.

From the Home Page, look for the section marked 'Topics'. Click on 'Engineering & Technology' to find collections of resources organized under these headings: Agriculture; Biomedical Engineering; Civil and Construction Engineering; Computer Technology; Electrical Engineering; Human Factors and Human Ecology; Mechanical Engineering; Production and Processing; Transportation; or 'Science' to find headings for Astronomy; Biology; Chemistry; Geosciences; Mathematics; Physics.

Gopher Jewels

```
gopher://cwis.usc.edu/11/Other_Gophers_and_Information_
Resources/Gopher_Jewels/
```

Gopher Jewels started as a voluntary project of David Riggins at the Texas Department of Commerce, but quickly grew to become one of the most popular tools for finding gopher resources by subject or application. The headings for science and technology include: Engineering and Industrial Applications; Health, Medical, and Disability; Natural Science including Mathematics; Technology Transfer and Grants.

Internet Resources Meta-Index

```
http://www.ncsa.uiuc.edu/SDG/Software/Mosaic/MetaIndex.html
```

The developers at NCSA, home of Mosaic, the popular graphical browser for the World-Wide Web, bring you this "a loosely categorized meta-index of the various resource directories and indices available on the Internet." It is regularly updated to include new resources.

University of California, San Diego Science and Engineering Library on WWW

```
http://scilib.ucsd.edu/

telnet://infopath.ucsd.edu

          login: infopath 'Library'; 'Library Branch
or        Collection';'Science and Engineering Library'
```

This is a beautiful and exceptionally well-designed 'Virtual Library' concentrating on Internet resources for research and teaching in Engineering and Physical Science (Chemistry, Computer Science, Mathematics, and Physics). It also offers several function hubs for jobs, grants, patents, and standards/specifications.

The telnet-accessible version is useful for those without WWW access.

WAIS Sources Organized by Subject

```
gopher://pinus.slu.se/11/wais-dbs/
```

This gopher resource not only lists WAIS sources by broad subject category, but also serves as a front-end for WAIS searching, allowing the user to enter search terms without leaving this site. Sci/tech categories are somewhat limited, but include Agriculture, Biology, Computers and Software, Environmental Sciences, Mathematics.

The WorldWide Web Virtual Library Subject Catalogue at CERN

```
http://info.cern.ch/hypertext/DataSources/bySubject/
Overview.html
```

```
http://info.cern.ch/hypertext/DataSources/bySubject/
LibraryofCongress.html
```

The Virtual Library Catalogue is developed and maintained by CERN, the Swiss physics research center that brought us the World-Wide Web. Their Virtual Libraries page includes headings for most every scientific and technical subjects: Aeronautics; Agriculture; Astronomy & Astrophysics; Aviation; Biosciences; Chemistry; Computing; Earth Science; Electronic Journals; Energy; Engineering; Environment; Fish; Forestry; Geography; Medicine; Meteorology; Oceanography; Physics; Statistics.

The Catalogue also includes a second organizational scheme based on the Library of Congress classification system. You may select this experimental section from the main Catalogue page, or point directly to it with the second URL provided above.

YAHOO (Yet Another Hierarchically O* Oracle)

```
http://www.yahoo.com/
```

Stanford researchers David Filo and Jerry Yang offer this selective, eclectic collection of science-related Internet resources. According to these developers, the asterisk in the YAHOO name may be filled in with various O-words.

The "Environment and Nature" section casts a wide net, covering Animal Rights to Weather and many topics in between. The science hierarchy covers aspects of many different fields, such as Acoustics, Astronomy, Biology, Chaos, Engineering, Forensics, Math, Paradoxes, Physics, and more. The resources included under each heading represent both the scholarly and the popular (and at times a merciless mix of the two).

Yahoo may lack some of the 'authority' of other subject-oriented resources described above, but it does provide an interesting selection of sites.... well worth a visit.

RESOURCE DISCOVERY TOOLS

When performing research in a library, one commonly relies on traditional retrieval tools such as catalogs, indexes, or abstracts to find needed information on a given subject. These tools typically allow searches across a variety of fields, including author, title, subject headings, or keyword. By performing a well-constructed search with one or more of these tools, you can feel confident that you have found most relevant published information on a given subject.

For the diverse, highly distributed resources 'published' over the Internet, there are no global catalogs or indexes available for performing a comprehensive subject search. The last few years, however, have seen the tremendous growth in resource discovery tools that retrieve and index information over certain parts of the Internet: they may search over a certain type of server (*e.g.*, Archie searches anonymous FTP archives, Veronica searches Gopherspace); over a certain subject (*e.g.*, the Peregrinator, a Web robot that ferrets out resources for mathematics or statistics); or over any resource, regardless of server protocol, that it can find. The databases used by the many Internet search tools may be built automatically, using robots that scour the Internet for resources, or manually, compiled from resource descriptions provided by information providers.

Because each search tool has distinct limits to the scope of its searches, it is usually necessary to try many different tools, trial-and-error style, to perform a 'comprehensive' search on the Internet. Each may turn up resources not retrieved in any other way.

The search tools that are useful for sci/tech researchers are listed below.

AliWeb (Archie-like Indexing for the Web)

`http://web.nexor.co.uk/public/aliweb/aliweb.html`

`http://www.cs.indiana.edu/aliweb/form.html (Indiana University mirror)`

AliWeb is resource discovery system for the World-Wide Web. It was developed in 1993 by Martijn Koster at NEXOR. The AliWeb database contains descriptions that have been written, in a standardized format, by various information providers on the World-Wide Web. Each AliWeb search returns a page of links matching the user's query, with results ranked in order of relevance. The database is fairly current (updated every day), but its scope is limited to resources for which providers have voluntarily submitted a description.

Archie

`telnet://archie.au`	`login: archie`
`telnet://archie.sura.net`	`login: archie`
`telnet://archie.ncu.edu.tw`	`login: archie`

Archie is one of the 'oldest' search tools on the Internet. It looks for matches to the user's search string in the directories and file names of all anonymous FTP sites on the Internet. The Archie sites listed above are just a few of the many mirror sites available around the world. For more information about Archie, see the Archie Fact Sheet in Section B of this guide.

ArchiePlex

`http://web.nexor.co.uk/archieplex-info/info.html`

ArchiePlex is an Archie gateway that allows Web users to look for files available by anonymous FTP. An ArchiePlex search returns a hypertext document with active links to the FTP hosts, directories and files that Archie has found.

The address referenced above is the primary site for ArchiePlex, but many mirror sites are available around the world.

Computer Science Technical Report Broker

`http://rd.cs.colorado.edu/brokers/cstech/query.html`

This is one broker built by the Harvest system described below.

According to the home page at the address listed above, "this broker covers more than 24,000 computer science technical reports from 300 sites around the world." Users search the index with WAIS, and search help is available online.

CUSI (Configurable Unified Search Engine)

`http://pubweb.nexor.co.uk/public/cusi.doc/list.html`
or `/about.html`

CUSI offers a convenient single interface for many searchable resources on the World-Wide Web. Developed by Martijn Koster at NEXOR, the CUSI service is available from the primary site at NEXOR, referenced above, and from many mirror sites worldwide.

Harvest Information Discovery and Access System

`http://rd.cs.colorado.edu/harvest/`

Harvest is actually a set of tools, the complete description of which spans some 15 pages at the site listed above. In a nutshell, Harvest is a flexible, customizable system for indexing and retrieving Internet resources for a particular topic or community. These tools then produce index servers, or brokers, to which a user inputs a search request. The resulting brokers vary quite a bit in their subject areas, search commands, searchable fields, etc.

A number of brokers created by Harvest system are available for searching on the World-Wide Web. One that may be of particular interest to readers of this book is the Computer Science Technical Report Broker described above.

HYTELNET

```
gopher://burrow.cl.msu.edu/11/internet/type/hytelnet
http://www.usask.ca/cgi-bin/hytelnet
http://galaxy.einet.net/hytelnet/HYTELNET.html
```

HYTELNET is the principal search tool for identifying and locating Telnet sites on the Internet. The 'HY' reflects the hypertext structure of this program. Most HYTELNET users load the software on their own Internet nodes, where the program acts both as a directory of all Telnet-accessible sites, and as a user-friendly interface from which one may select Telnet connections.

You may also search HYTELNET from one of the public sites specified above. On both the Gopher and the Einet WWW gateways, you may either perform keyword searches of all entries in the HYTELNET database, or browse through selections on the menus.

Jughead List

```
gopher://honor.uc.wlu.edu/11/gophers/jugheads
```

Jughead is the search and retrieval system for limited portions of Gopherspace, and this site provides an extensive list of Jugheads, each of which searches over a particular group of Gophers (*e.g.* all Biology gophers, all Japanese Gophers).

JumpStation

```
http://www.stir.ac.uk/jsbin/js
```

Developed in late 1993 by Jonathon Fletcher at Stirling University, JumpStation is an experimental Web robot. The scope of JumpStation searches is limited to the text in an HTML document's title or header (*i.e.*, text between the <ti> </ti> or <h1></h1> tags). JumpStation can also perform searches across URL's, making it a useful tool for re-discovering sites when only a partial URL is known.

MathSearch

```
http://www.math.usyd.edu.au:8000/MathSearch.html
```

This index, covering over 8000 documents in math and statistics, is the product of a particular Web Robot, the Peregrinator (see 'WWW List of Robots' site for more information). Because this robot is on a tight leash (it wanders select Internet servers specified upfront), it may miss some relevant resources.

Veronica

`gopher://veronica.scs.unr.edu/11/veronica`

Veronica is a tool that searches all of gopherspace for the users' search string. Veronica returns a customized gopher menu that includes all matching menu items and/or all directories titles. For more information about Veronica, see the Veronica Fact Sheet in Section B of this guide.

WAIS Directory of Servers

`telnet://quake.think.com  login:wais`

`http://www.wais.com/directory-of-servers.html`

The Directory of Servers is a catalog of some 500 or more databases and files searchable by WAIS (ie., all the '.src' files known to WAIS.) When you run a keyword search in the Directory of Servers, you get a listing of sources with a title or description matching your search terms. You may then read those descriptions and select which source(s) to search for actual documents, images, and all the other resources you can retrieve with WAIS.

Remember that the Directory of Servers is just a catalog of all sources, describing what each source has to offer. The Directory of Servers does not provide access to the actual content of those sources.

For more information on searching WAIS, see the WAIS Fact Sheet in Section B of this guide. Also see the 'WAIS Sources Organized by Subject' section above.

WAIS Sources via Gopher

`gopher://liberty.uc.wlu.edu/11/internet/indexsearches/`
`inetsearches/`

Here Washington & Lee has pulled together a variety of links to gopher interfaces to WAIS, making WAIS searching and retrieval a little easier and more convenient.

W3 Search Engines

`http://cuiwww.unige.ch/meta-index.html`

This home page provides links to many of the search tools available on the Internet. For each tool listed, the page provides a brief description of the tool, a direct link to the ultimate server where the tool actually resides, and a user-input form to allow convenient searching right from the W3 Search Engines home page. This is a great site for trying the same search on different search tools, and comparing/contrasting the results.

WWW List of Robots

`http://web.nexor.co.uk/mak/doc/robots/active.html or /`
`robots.html`

Behind the scenes, a number of tools are quietly scouring the World-Wide Web for resources, indexing documents to varying extents, and delivering the information back to publicly-accessible servers where users can pose their queries to the index. In some cases, the tools may act as agents, going back out to the Web to find information known to be relevant to the user's query.

These tools are variously known as 'robots', 'spiders', 'wanderers', or 'worms'. One doesn't need to be an arachnophobe to feel dizzied by the many possible tools and their differing capabilities. Those interested in trying one or more of these tools may visit this home page at Nexor, which provides a brief description of each tool and direct links to their servers. Also note that some of the better-known robots, spiders, etc. are also listed separately in this section.

For a more detailed discussion of Web robots, spiders and wanderers, please see the 'WWW Robots' Fact Sheet in Section B of this guide.

WWW Catalog at CUI

`http://cuiwww.unige.ch/w3catalog`

This catalog is manually developed from a selective combination of other directories, resources and indexing tools on the World-Wide Web.

WWW WebCrawler Index

`http://webcrawler.cs.washington.edu/WebCrawler/`
`WebQuery.html`

WebCrawler is a robot developed by Brian Pinkerton at the University of Washington. It produces an index compiled from over 40,000 documents on 4000 WWW servers worldwide. It builds the index from the full text of the home pages it scours, as well as the home pages it finds by following the links at the original starting point. Based on entirely empirical evidence, the WebCrawler is a personal favorite.

WWW Worm

`http://www.cs.colorado.edu/home/mcbryan/WWWW.html`

This WWW robot, maintained by Oliver McBryan, identifies and indexes WWW documents by tracking URL's that are cited in WWW documents. In other words, WWWW cross-references the URL's, or anchors, with the surrounding text in which it is cited, thereby giving immediate context to the linked resource specified by the URL. The WWWW index includes 300,000 multimedia resources, including text documents, images, MPEG movies, and more.

WWW Worm's citation-based indexing system creates a highly effective retrieval tool. That's the feeling of the Web community, anyway—they awarded this worm as 'Best Navigational Aid' in the 'Best of the Web '94' competition.

Lycos WWW search engine at Carnegie Mellon University

`http://lycos.cs.cmu.edu/`

Lycos is a robot and search system for select resources on the WorldWide Web (primarily textual documents served by HTTP, Gopher, and FTP protocols). Lycos bases its indexing on summaries of Web pages that include title, headings and subheadings, links or anchors, the highest weighted words, and the first 20 lines of the document. The robot searches the Web daily, and the database currently contains 2.5 million URL's.

Developed by Carnegie Mellon's Center for Machine Translation, Lycos is experimental at this point, and it is best to read the introductory information about the system at the home page listed above.

And for those not fully conversant in arachnid-ology, the home page is quick to point out that Lycos is short for the speedy, nocturnal Lycosidae family of spiders, known to catch their prey by pursuit rather than by web.

PROSPECTING FOR POINTERS

Professional journals, bulletins and newsletters

Prospecting for pointers in the professional publications of your own field will steer you to sites deemed to be higher-quality and relevant by your own colleagues. For example, the American Geophysical Union's weekly *EOS* newsletter commonly describes searchable databases and other sites of interest to the geophysics community. The American Institute of Physics' monthly *Computers in Physics* runs a regular column "Internet Corner," featuring physics-related Internet services and directions for accessing them. The library-related journals *Database*, *Online*, and *College and Research Libraries News* also feature reviews of Internet resources for diverse subject areas, including science and technology.

You may find such articles by performing a 'traditional' literature search in a relevant index. For example, a recent search in *Current Contents* [1] using the keywords 'Internet' and 'Engineering' located an article[2] describing resources for industrial engineers, including FTP, Gopher and WWW sites, discussion groups, and more. Index and abstracting services covering many other fields of science or technology, such as *Biological Abstracts*, *Chemical Abstracts*, *Index Medicus*, and others, are sure to include articles about Internet resources for the field.

Electronic Discussion Groups

Discussion groups for most every field of science and technology are now available on the Internet, and these forums often post information about new or useful network sites of interest to the group. For more information about identifying and joining discussion groups in your area of interest, see the 'Conferencing and Collaboration' module in Section A of this guide.

NewJour-L@ccat.sas.upen.edu

mail to:

```
listserv@ccat.sas.upen.edu
```

message:

```
subscribe newjour-l YourName
```

Association of Research Libraries' distribution list posts announcements about new or revised electronic journals or newsletters in planning or production.

New-List@vm1.nodak.edu

mail to

```
listserv@vm1.nodak.edu
```

message:

```
subscribe new-list YourName
```

A clearinghouse for announcements of new electronic discussion groups.

---

1 In this case, the version searched was *Current Contents* on LUIS, the online public catalog system for Florida's State University System Libraries.

2 Mathieu, Richard G. "The Internet: Information Resources for Industrial Engineers" *Industrial Engineering* (January 1995): 49-52.

Section D: Selected Readings

The following articles, reports, and other documents are particularly insightful and helpful for understanding how the Internet is being used in the many disciplines of science and technology, and its impact on the processes of scientific investigation, information/resource sharing, and communication.

Past Developments and Future Trends

Bowman, C. Mic *et al.* "Scalable Internet Resource Discovery: Research Problems and Approaches" *Communications of the ACM* vol. 37, no. 8 (August 1994): 98-107, 114.

Cerf, Vint, as told to Bernard Aboba, "How the Internet Came To Be", Internic InfoGuide:

```
gopher://is.internic.net/00/infoguide/about-internet/his-
tory/aboba-cerf
```

Cerf, Vint. Networks. "Networks." *Scientific American* vol. 265, no. 3 (September 1991): 72-81.

Claffy, Kimberly C., Hans-Werner Braun, and George C. Polyzos. "Tracking Long-Term Growth of the NSFNET". *Communications of the ACM* vol. 37, no. 8 (August 1994): 34-45.

Dertouzos, Michael L. "Communications, Computers and Networks". *Scientific American* vol. 265, no. 3 (September 1991): 62-69.

Hoke, Franklin. "Scientists Predict Internet Will Revolutionize Research" *The Scientist* vol. 8, no. 9 (May 2, 1994): 1-7.

Kahn, Robert E. "The Role of the Government in the Evolution of the Internet" *Communications of the ACM* vol. 37, no. 8 (August 1994): 15-19.

Kantrowitz, Barbara and Adam Rogers. "The Birth of the Internet." *Newsweek* (August 8, 1994): 56-57.

Pool, Robert. "Beyond Databases and E-Mail" *Science* vol. 261, no. 5123 (August 13, 1993): 841, 843.

Sproull, Lee and Sara Keiser. "Computers, Networks, and Work." *Scientific American* vol. 265, no. 3 (September 1991): 116-123.

Watson, Richard. "Creating and Sustaining a Community of Global Scholars" *MIS Quarterly* (September 1994): 225-231.

Weiser, Mark. "The Computer for the 21st Century" *Scientific American* vol. 265, no. 3 (September 1991): 94-104.

Wulf, William. "The Collaboratory Opportunity" *Science* vol. 261, no. 5123 (August 13, 1994): 854-855.

Applications of Today: Some Examples and Resources

Anderson, Christopher. *"Cyberspace Offers Chance to Do 'Virtually' Real Science'*, *Science* vol. 264 (May 13, 1994): 900-901.

Bradley, Raymond S., Linda G. Ahern, and Frank T. Keimig. "A Computer-based Atlas of Global Instrumental Climate Data." *Bulletin of the American Meteorological Society* vol. 75, no. 1 (1994): 35-41.

Chandler, David L. "Jupiter-comet crash researchers' invaluable tool: Internet" *Miami Herald*, (July 31, 1994), 6M.

Clauer, C. R. et al. "New Project to Support Scientific Collaboration Electronically" *EOS* vol. 75, no. 26 (June 28, 1994): 289, 295, 298.

Commission on Physical Sciences, Mathematics, & Applications Staff, National Research Council. *National Collaboratories: Applying Information Technology For Scientific Research.* Washington D.C.: National Academy Press, 1993.

Cox, Jennifer and Mohamed Taleb. "Images on the Internet." Database (August 1994): 18-26.

Doherty, Michael E. Jr. "MOO as Tool, MOO as Realm" *Computer-Mediated Communication Magazine* vol. 1, no. 7 (November 1, 1994).
    `http://www.rpi.edu/~decemj/cmc/mag/archive.html`

Green, David G. "Databasing Diversity — A Distributed, Public-domain Approach" *Taxon* vol. 43, no. 1 (February 1994): 51-62.

Eriksson, Hans. "MBONE: The Multicast Backbone". *Communications of the ACM* vol. 37, no. 8 (August 1994): 34-59.

Heller, Stephen R. "Analytical Chemistry Resources on the Internet." *Trends in Analytical Chemistry* vol. 13, no. 1 (1994): 7-12.

Hoke, Franklin. "New Internet Capabilities Fueling Innovative Science" *The Scientist* vol. 8, no. 10 (May 16, 1994): 1-6.

Macedonia, Michael R. and Donald P. Brutzman. "MBone Provides Audio and Video Across the Internet". *IEEE Computer*, vol. 27, no. 4. April 1994, pp. 30-36. Available online as:
    `file://taurus.cs.nps.navy.mil/pub/mbmg/mbone.html.`

Marchionini, Gary, Diane Barlow, and Linda Hill. "Extending Retrieval Strategies to Networked Environments: Old Ways, New Ways, and a Critical Look at WAIS" *Journal of the American Society for Information Science* vol 45, no. 8 (1994): 561-564.

Martin, Nadia J., Tracy Primich, and Ruth A. Riley "Accessing Genetics Databases" *Database* (February 1994): 51-58.

Mathieu, Richard G. "The Internet: Information Resources for Industrial Engineers" *Industrial Engineering* (January 1995): 49-52.

Maxymuk, John "Science Resources on the Internet" *The Reference Librarian* no. 41/42 (1994): 81-98.

Menke, William, Paul Friberg, Arthur Lerner-Lam et al. "Sharing Data Over Internet with the Lamont View-Server System" *EOS* vol. 72, no. 38, (September 17, 1991): 409, 413.

Mindell, David A. "Images From the Deep." *Byte* (June 1990): 256-260.

Pool, Robert. "Networking the Worm" *Science* vol. 261, no. 5123 (August 13, 1993): 842.

Roamowicz, Barbara, *et al.* "Accessing Northern California Earthquake Data Via Internet" *EOS*, vol. 75, no. 23 (June 7, 1994): 257, 259.

Starr, Susan S. "Evaluating Physical Science Reference Sources on the Internet." *The Reference Librarian* no. 41/42 (1994): 261-273.

Thoen, Bill "Access the Electronic Highway for a World of Data" *GIS World* vol. 7, no. 2 (February 1994): 46-49.

Waldrop, M. Mitchell. "Software Agents Prepare to Sift the Riches of Cyberspace" *Science* vol. 265 (August 12, 1994): 882-883.

Weiss, Aaron "Stretching the MBone" *Internet World vol. 6, no. 3* (March 1995): 38-41. Also available online as:

```
gopher://gopher.enews.com:2100/00/magazines/alphabetic/gl/
internet_world/Archive/030195.3
```

Electronic Publishing

Bailey, Charles W. Jr. "Electronic Publishing on Networks: A Selective Bibliography of Recent Works" *The Public-Access Computer Systems Review* vol. 3, no. 2 (1992): 13-20.

Clement, Gail P. "Evolution of A Species: Science Journals Published on the Internet." *Database* vol. 17, no. 5 (1994): 44-54. Available online as:

```
gopher://online.lib.uic.edu/0F-1%3a1632%3a%2a0Clement
```

Harnard, Stevan "Scholarly Skywriting and the Prepublication Continuum of Scientific Inquiry" *Psychological Science* vol. 1 (1990): 342-343

Keyani, Andrea "The Online Journal of Current Clinical Trials: An Innovation in Electronic Journal Publishing" *Database* (February 1993): 14-23.

Lesk, Michael E. "Electronic Chemical Journals" *Analytical Chemistry* vol. 66, no. 14 (July 15, 1994): 747-755.

Lucier, Richard E. and Robert C. Badger "Red Sage Project" *The Serials Librarian* vol. 24, no. 3/4 (1994): 129-134.

Okerson, Ann. "The Electronic Journal: What, Whence, and When?" *The Public Access Computer Systems Review* vol. 2, no. 1 (1991): 5-24.

Rodgers, David L. "Scholarly Journals in 2020" *The Serials Librarian* vol. 24, no. 3/4 (1994): 73-76.

Schaffner, Ann C. "The Future of Scientific Journals: Lessons from the Past" *Information Technology and Libraries* vol. 13, no. 4 (December 1994): 239-247.

Schauder, D. "Electronic Publishing of Professional Articles: Attitudes of Academics and Implications for the Scholarly Communication Industry." *Journal of the American Society for Information Science* vol. 45, no. 2 (March 1994): 73-100.

Stix, Gary. "The Speed of Write" *Scientific American* (December 1994): 106-111.

Taubes, Gary. "Peer Review in Cyberspace" *Science* vol. 266, no. 5187 (November 11, 1994): 967.

Van Steenberg, Michael E. "NASA STELAR Experiment" *The Serials Librarian* vol. 24, no. 3/4 (1994): 135-51.

The Second Revolution: Science, Technology, and the World-Wide Web

Aiton, James F. "The World-Wide Web: An Interface Between Research and Teaching in Bioinformatics" *Disease Markers* vol. 12 (1994): 3-10.

Barry, Jeff. "The HyperText Markup Language (HTML) and the WorldWide Web: Raising ASCII Text to a New Level of Useability." *PACS Review* vol. 5, no. 5 (1994).

Available by email to
```
listserv@uhupvml.uh.edu; message 'get barry prv5n5 f=mail'.
```

Berners-Lee, Tim, Robert Cailliau, Ari Luotenen, et. al "The World-Wide Web." *Communications of the ACM* vol. 37, no. 8 (August 1994): 76-82.

Bush, Vanevar. "As We May Think." *Atlantic Monthly* (1945): 101-108.

O'Dea, Christopher P. "Hubble Space Telescope Via the Web" *International Journal of Modern Physics* vol. 5, no. 5 (1994): 811-816.

Pasian, Fabio and Riccardo Smareglia "WWW Access to Astronomical Archives and Databases" *International Journal of Modern Physics* vol. 5, no. 5 (1994): 817-830.

Rzepa, Henry S., Benjamin J. Whitaker and Mark J. Winter. " Chemical Applications of the World-Wide-Web System." *Jour. Chem. Soc. Chem. Comm.* (1994): 1907-1910.

Schatz, Bruce R. and Jospeh B. Hardin, "NCSA Mosaic and the World Wide Web" Global Hypermedia Protocols for the Internet." *Science* vol. 265 (August 12, 1994): 895-901.

Section E: Presentation Slides

The slides on the following pages may be used to make transparencies for use on an overhead projector. They are printed on perforated pages for easy removal. Each slide is numbered to correspond with the numbered projector icon printed in Section A. The slide should be displayed when you reach its corresponding number in the text.

The diskettes at the back of the book contain these same slides in color.

1. Science and Technology on the Internet
2. What We'll Cover

Conferencing and Collaboration

3. What's Available
4. Applications and Access Methods
5. Quality and Value

Finding Colleagues

6. Strategies
7. Use Combination of Approaches
8. Tools

News and Current Information

9. What's Available
10. How to Access It
11. Quality and Value
12. How to Find It

Reference Tools

13. Reference Tools
14. Access Methods
15. Six Criteria for Evaluation

Searching the Literature

16. What's Available
17. Literature Search Sources: Quality and Value
18. Access Methods

Electronic Publishing

19. What's Available
20. Quality and Value
21. Access Methods

Data and Resource Sharing

22. What's Available
23. Methods for Sharing Data and Resources
24. Quality and Value

SCIENCE AND TECHNOLOGY ON THE INTERNET

Science and Technology on the Internet

What We'll Cover

- Conferencing and collaboration
- Finding colleagues
- News and current information
- Reference tools
- Searching the literature
- Electronic publishing
- Sharing data and resources

Conferencing & Collaboration: What's Available

- Electronic discussion groups
- Virtual reality environments
- Video conferences and multicasting

Applications and Access Methods

- **Electronic Discussion Groups**
  - Mailserver lists
  - Network/Usenet Newsgroups
- **Virtual environments**
  - IRC
  - MOO

Applications and
Access Methods (cont.)

- Videoconferencing
  - MBone
  - C U C Me
- Archives, FAQs, etc.
  - FTP
  - Gopher
  - WAIS
  - WWW

Quality and Value

- In moderated forums, the moderator

  - screens participants

  - filters contributions for relevance

  - may lead or referee discussion

- In unmoderated forums

  - self-regulating behavior

  - voluntary agreement to abide by group's aims and rules

Finding Colleagues: Strategies

- No all-inclusive directory

- Criteria for assessing tools and services

  - What's the scope? How complete?

  - How current? How accurate?

  - How much information is given?

  - How easy to use?

  - Report given for unsuccessful search?

Use Combination of Approaches

- Email to discussion group
- Telnet to directory service (directly or via Gopher or WWW)
- Gopher-based searching tools
- WAIS-based directories
- WWW indexing tools (e.g., WebCrawler)

Tools

- Organizational phonebooks and directories

- Professional directories

- Discussion group subscriber lists and archives

- Distributed Internet directory services

News & Current Information: What's Available

- News of natural phenomena

- Research developments

- Conference announcements, calls for papers

- Funding, employment, fellowship opportunities

- RFPs

News & Current Information: What's Available (cont.)

- Updates to government policies & regulations
- Latest versions of proposal forms
- Press releases
- Product announcements

How to Access It

- ➤ Mailserver discussion groups, Network Newsgroups
- ➤ Announcements posted to an organization's Internet server
- ➤ Digital broadcasting tools
  - ➤ IRC
  - ➤ MOO
  - ➤ MBone

Quality and Value

- ➢ Trade-off between timeliness and reliability
- ➢ Consider the source
- ➢ Heed disclaimers
- ➢ Note contact address for site administrator

How to Find It

- **Step 1** Identify organization or publisher disseminating the news

- **Step 2** Locate its address using good hubs (starting sites)

- **Other approaches to try:**
  - Use tools such as Veronica or WWW Spider
  - Check archives of likely discussion group

Reference Tools:
What's Available

➤ **Standards and specifications**

➤ **Facts and data**

➤ **Encyclopedias, dictionaries, atlases**

➤ **Scales of measurements**

➤ **Reference textbooks**

➤ **Supply catalogs**

➤ **Bibliographies, indexes and abstracts**

Access Methods

> ➤ Gopher

> ➤ WWW

> ➤ Finger

> ➤ WAIS (limited)

> ➤ Mailservers (minimal)

> ➤ FTP (minimal)

Six Criteria for Evaluation

1. **Purpose:** Does it fulfill stated purpose?

2. **Authority:** Author's credentials?

3. **Scope:** Depth? Breadth? How does it compare with others?

4. **Audience:** Who? What level of expertise expected?

5. **Cost:** User fees? Software/hardware/ telecommunications expenses?

6. **Format:** How easy, convenient, effective to use?

Searching the Literature: What's Available

- ➤ Abstract and indexing services
- ➤ Full-text and graphics databanks
- ➤ Specialized bibliographies and library catalogs
- ➤ Current awareness services
- ➤ Document delivery services

Quality and Value

➤ **Scope**
➤ **Selection process for material included**
➤ **Format**

  ➤ structure

  ➤ number and types of fields

  ➤ search and display features

➤ **Currency**

Access Methods

➢ Email
➢ Telnet
➢ FTP
➢ WAIS
➢ Gopher
➢ WWW

Electronic Publishing: What's Available

- Parallel publishing
- Supplementary material
- Grey literature sources
- Network science journals
- E-prints

Quality and Value

- **Authority, reliability**

  (does not apply to e-prints)

  - Peer-reviewed?
  - Editorial process?

- **Archives**

  - How permanent?
  - How complete?
  - How current?

Quality and Value (cont.)

- ➤ Access
  - ➤ Currently spotty
  - ➤ Slow to be covered by traditional indexing/ abstracting services

Access Methods

- ➤ Email (with MIME)
- ➤ FTP
- ➤ Gopher
- ➤ WAIS (text files only)
- ➤ WWW

Sharing Data and Resources: What's Available

- Data directories
- Central databanks
- Individual data sets
- Software repositories
- Multimedia libraries

Methods for Sharing Data and Resources

- Email
- Mailservers
- Telnet
- FTP
- WAIS
- Gopher
- Specialized software agents

Quality and Value

➤ Made available "as is"

➤ Investigator is responsible for validity, accuracy

➤ Moderator may screen for relevance

➤ Curator may screen for consistency, conformance to standards

➤ Executable files are vulnerable to viruses

Presentation Slides

Science and Technology on the Internet

What We'll Cover

- Conferencing and collaboration
- Finding colleagues
- News and current information
- Reference tools
- Searching the literature
- Electronic publishing
- Sharing data and resources

Conferencing & Collaboration: What's Available

- Electronic discussion groups
- Virtual reality environments
- Video conferences and multicasting

Applications and Access Methods

- Electronic Discussion Groups
  - Mailserver lists
  - Network/Usenet Newsgroups
- Virtual environments
  - IRC
  - MOO

Applications and Access Methods (cont.)

- Videoconferencing
  - MBone
  - C U C Me
- Archives, FAQs, etc.
  - FTP
  - Gopher
  - WAIS
  - WWW

Quality and Value

- In moderated forums, the moderator
  - screens participants
  - filters contributions for relevance
  - may lead or referee discussion
- In unmoderated forums
  - self-regulating behavior
  - voluntary agreement to abide by group's aims and rules

Finding Colleagues: Strategies

- No all-inclusive directory
- Criteria for assessing tools and services
  - What's the scope? How complete?
  - How current? How accurate?
  - How much information is given?
  - How easy to use?
  - Report given for unsuccessful search?

Use Combination of Approaches

- Email to discussion group
- Telnet to directory service
  (directly or via Gopher or WWW)
- Gopher-based searching tools
- WAIS-based directories
- WWW indexing tools (e.g., WebCrawler)

Tools

- Organizational phonebooks and directories
- Professional directories
- Discussion group subscriber lists and archives
- Distributed Internet directory services

News & Current Information: What's Available

- News of natural phenomena
- Research developments
- Conference announcements, calls for papers
- Funding, employment, fellowship opportunities
- RFPs

News & Current Information: What's Available (cont.)

- Updates to government policies & regulations
- Latest versions of proposal forms
- Press releases
- Product announcements

How to Access It

- Mailserver discussion groups, Network Newsgroups
- Announcements posted to an organization's Internet server
- Digital broadcasting tools
  - IRC
  - MOO
  - MBone

Quality and Value

- ➤ Trade-off between timeliness and reliability
- ➤ Consider the source
- ➤ Heed disclaimers
- ➤ Note contact address for site administrator

 How to Find It

- ➤ Step 1　Identify organization or publisher disseminating the news
- ➤ Step 2　Locate its address using good hubs (starting sites)
- ➤ Other approaches to try:
  - ➤ Use tools such as Veronica or WWW Spider
  - ➤ Check archives of likely discussion group

 Reference Tools: What's Available

- ➤ Standards and specifications
- ➤ Facts and data
- ➤ Encyclopedias, dictionaries, atlases
- ➤ Scales of measurements
- ➤ Reference textbooks
- ➤ Supply catalogs
- ➤ Bibliographies, indexes and abstracts

 Access Methods

- ➤ Gopher
- ➤ WWW
- ➤ Finger
- ➤ WAIS (limited)
- ➤ Mailservers (minimal)
- ➤ FTP (minimal)

 Six Criteria for Evaluation

1. Purpose: Does it fulfill stated purpose?
2. Authority: Author's credentials?
3. Scope: Depth? Breadth? How does it compare with others?
4. Audience: Who? What level of expertise expected?
5. Cost: User fees? Software/hardware/ telecommunications expenses?
6. Format: How easy, convenient, effective to use?

 Searching the Literature: What's Available

- ➤ Abstract and indexing services
- ➤ Full-text and graphics databanks
- ➤ Specialized bibliographies and library catalogs
- ➤ Current awareness services
- ➤ Document delivery services

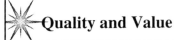

Quality and Value

- Scope
- Selection process for material included
- Format
  - structure
  - number and types of fields
  - search and display features
- Currency

Access Methods

- Email
- Telnet
- FTP
- WAIS
- Gopher
- WWW

Electronic Publishing: What's Available

- Parallel publishing
- Supplementary material
- Grey literature sources
- Network science journals
- E-prints

Quality and Value

- Authority, reliability
  (does not apply to e-prints)
  - Peer-reviewed?
  - Editorial process?
- Archives
  - How permanent?
  - How complete?
  - How current?

Quality and Value (cont.)

- Access
  - Currently spotty
  - Slow to be covered by traditional indexing/ abstracting services

Access Methods

- Email (with MIME)
- FTP
- Gopher
- WAIS (text files only)
- WWW

Sharing Data and Resources: What's Available

- Data directories
- Central databanks
- Individual data sets
- Software repositories
- Multimedia libraries

Methods for Sharing Data and Resources

- Email
- Mailservers
- Telnet
- FTP
- WAIS
- Gopher
- Specialized software agents

Quality and Value

- Made available "as is"
- Investigator is responsible for validity, accuracy
- Moderator may screen for relevance
- Curator may screen for consistency, conformance to standards
- Executable files are vulnerable to viruses

INDEX

Peer-reviewed journals on the Internet xviii
Peregrinator, example of WWW Robot 190
Phonebook hubs
    exercises in using 34-35, 36-37, 38-39
    for finding email directories 32
Physics hub 203
PINE
    example of mail system that supports MIME 175
    exercise in using 64-67
Place, finding information about a remote
    (exercise) 57-58
Preprint, making yours available as an e-print
    (exercise) 115-117
Preprints, electronic 104
Professional directories
    as a tool for finding email addresses 31-32
    use to find an email address (exercises) 36-37
Professional journals, use to find out about good sites 215
Publications, electronic 101-126
Publisher's WWW site, search a (exercise) 94-96
Publishing, Internet developments in electronic 101-104
Publishing on the Internet, instructional module on
    101-126
Publishing simultaneously in print and online 102
Publishing supplementary material electronically 102

Quakeline, search for bibliographic information in
    (exercise) 83-86
Quality and value of (how to assess)
    databank services 132
    electronic publications 104-105
    Internet conferencing services 8-9
    literature search services 80
    news and current information resources 46-47
    reference tools 61
    services for finding email address 29-30
Readings 217-222
Reference tools
    how to assess the quality of 61
    instructional module on 59-76
    Internet developments in 59-60
Relevance of the Internet to scientists' work xvii-xix
Research project, find out about a current (exercise) 50
Resource discovery tools
    list of 209-215
    WWW Robots as examples of 190-191
Resource sharing, instructional module on 127-145
Retrieve an article in an unfamiliar format
    (exercise) 121-122
Retrieving and downloading files
    (exercises) 109-111, 112-114
Reviews of literature, search the full-text of
    (exercise) 97-99

Save information retrieved via WWW to a local file
    (exercise) 76
Scholarly Communications Project 129
Scholarly Discussion Groups, Directory of 207

Scholarly Societies hub 38, 48, 205
Science journals published only on the Internet 103
Science, technology, and the Internet (background) xvii-xix
Scientific visualizations hub 205
Scientists and the Internet xvii-xix
Sea Turtles bibliography, example of specialized
    database 79
Searching and retrieving the literature 77-99
SGML 164
Sharing data and resources
    instructional module on 127-145
    Internet developments in 127-132
Simple Mail Transfer Protocol 151, 175
Simple WAIS (SWAIS), used in exercise 97-99
SMTP 151, 175
Societies and associations hub 205
Software agents for accessing databank services 132
Software, find and retrieve (exercises) 135-136, 137-139
Software repositories 129-130
Solinet Gopher, use to access Virtual Reference Desks 61
Specialized bibliographic databases 78
Standard Generalized Markup Language 164
Standards hub 205
Standards, search for using WAIS-via-Gopher
    (exercise) 72-73
Starting a Usenet Newsgroup (exercise) 21-22
Starting your search, good sites for 195-216
Statistics hubs 202
STELAR project 79, 82
STIS, exercise in using 55-56
Strategies for finding
    bibliographic resources 81-82
    email addresses 29-30
Subject approach to finding Internet resources 195-216
Subject directories of Internet resources 205-209
Subject guides to the Internet 197-200
Subject hubs
    list of 200-202
    use for finding reference sources 61
Subject-organized resources 205-209
Subscribe to a listserv, exercise 14-16
Subscribe to a Usenet Newsgroup, exercise 17-20
Subscriber lists of discussion groups, use for finding email
    addresses 32
Supersites *see* Hubs
Supplementary material published electronically 102, 129
SWAIS (simple WAIS)
    as a version of WAIS 188
    used in exercise 97-99

Table of contents of journal, find a (exercise) 92-93
Tables of contents of journals, services that provide 79
Technical report services 102-103
Telnet
    as a means for transmitting data 131
    exercises using 36-37, 38, 41-43, 83-86
    fact sheet on 181-182
    use to search a bibliography (exercises) 83-86, 87-89